Advances in Peircean Mathematics

Peirceana

Edited by
Francesco Bellucci and Ahti-Veikko Pietarinen

Volume 7

Advances in Peircean Mathematics

—

The Colombian School

Edited by
Fernando Zalamea

DE GRUYTER

ISBN 978-3-11-152170-1
e-ISBN (PDF) 978-3-11-071763-1
e-ISBN (EPUB) 978-3-11-071771-6
ISSN 2698-7155

Library of Congress Control Number: 2022945932

Bibliographic information published by the Deutsche Nationalbibliothek
The Deutsche Nationalbibliothek lists this publication in the Deutsche Nationalbibliografie; detailed bibliographic data are available on the Internet at http://dnb.dnb.de.

© 2024 Walter de Gruyter GmbH, Berlin/Boston
This volume is text- and page-identical with the hardback published in 2023.
Typesetting: Jukka Nikulainen

www.degruyter.com

May some future student go over this ground again, and have the leisure to give his results to the world.

— Charles Sanders Peirce (1891) *Collected Papers* 6.34.

Contents

Dedication —— IX

Acknowledgments —— XI

Fernando Zalamea
Introduction —— XIII

Gustavo Arengas
1   **Category Theory Variations and Proofs of Peirce's Pragmaticist Maxim —— 1**
1.1   Unifying Correspondences —— 2
1.2   Signs and ∞-Categories —— 7
1.3   Unlimited Semiosis and (co)Limits —— 15
1.4   Pragmatic Maxim and Monads —— 21
1.5   Pragmatic Maxim and Descent Theory —— 29
1.6   Determination Process and Adjoint Strings —— 38
1.7   Further Venues —— 51

Francisco Vargas
2   **A Full Model for Peirce's Continuum —— 55**
2.1   Peirce's Riddles on Continuity —— 57
2.2   A Set Theoretic Reconstruction —— 64
2.3   A Mereological View —— 69
2.4   A Kripke Model View —— 72
2.5   From the Infinitely Small to the Infinitely Large —— 76
2.6   From the Line to the Plane and Beyond —— 79
2.7   An Algebraic View —— 81
2.8   A Geometric View —— 85
2.9   Infinitesimal Calculus: an Alternative Approach —— 86
2.10   Further Paths —— 102

Arnold Oostra
3   **Intuitionistic and Geometrical Extensions of Peirce's Existential Graphs —— 105**
3.1   Peirce's Existential Graphs —— 105
3.2   Intuitionistic Existential Graphs —— 122
3.3   Existential Graphs on Surfaces —— 162

Fernando Zalamea
4   **Around Arengas, Vargas, and Oostra Models for Peirce's Thought** —— **181**
4.1   Arengas Models —— 183
4.2   Vargas Models —— 186
4.3   Oostra Models —— 188
4.4   Hugueth Drawings —— 190
4.5   Towards a Mathematical Architectonics of Peirce's Architectonics —— 194

Peirce Bibliography —— **199**

Secondary Bibliography —— **201**

Name Index —— **209**

Keyword Index —— **211**

Dedication

We dedicate this volume to the loving memory of our friends and colleagues Yuri Poveda (1968–2021), who first offered a finite linear axiomatization of Peirce's existential graphs, and William James McCurdy (1947–2021), a brilliant scholar in the Mathematics of Peirce, who was accomplishing a profound foundational program in the combinatorial topology of Peirce's logic of relations.

Acknowledgments

Evolving slowly over the last 20 years, this volume has benefited from many hands. In particular, the visits to Bogotá of dear friends and Peirce scholars helped much to foster the investigations included in this volume. Thanks are due to our visitors (some of them frequent) Shigeyuki Atarashi, Jeff Downard, Nicholas Guardiano, Jérôme Havenel, Catherine Legg, Giovanni Maddalena, Rosa Mayorga, William James McCurdy, Matthew Moore, Jaime Nubiola, Ahti-Veikko Pietarinen, John Sowa. On another hand the *Charles S. Peirce International Centennial Congress* (Lowell, 2014) set a high standard that we tried to follow in this volume.

Special thanks are due to Gustavo Arengas for solving difficult TEX impasses, to Francisco Vargas for many improvements in the general and particular outlook of the volume, and to Arnold Oostra for his continuing support and strong immersion in the edition of all chapters. The three of them, Arnold, Francisco, Gustavo, produced extremely precise revisions, beyond all duties, of the entire book. Two readers for De Gruyter offered amazingly careful suggestions and comments, which enhanced and helped to correct the final product: many thanks to them. Finally, many thanks also to Jukka Nikulainen for the extraordinary typesetting of the final manuscript, and to Mara Weber for the overall edition of the volume.

Fernando Zalamea
Introduction

The Colombian School

In Latin America, and more generally in the Hispanic World, pragmatism came to our countries mainly through William James and John Dewey (particularly, through his influence in education). Studies in Spanish on Peirce were scarce and superficial until the end of the 20th century. The situation in Colombia follows that pattern. The slow development of a *Colombian School* on Peirce began with Fernando Zalamea's Peircean Seminars (1996–1999) given at the Universidad Nacional de Colombia, where an important number of young scholars gathered. Zalamea centered his research on existential graphs and the continuum (1997–2001), converging in the first international monograph (2012) devoted to the mathematical entanglements of both the continuum and the graphs.

Meanwhile, Zalamea's influence spread out (11 Thesis on Peirce), and students and colleagues began their own path, with the appearances of Eugenio Andrade (1999–2000) in biological thought, Douglas Niño (2000) in semiotics, and Arnold Oostra (2001–2004) in logic. Zalamea donated his Peircean books and microfilms to the Universidad Nacional, in order to create the *Acervo Peirceano* (1999), possibly the largest collection of specialized material on Peirce in Latin America.

The main Colombian Peirce scholars continued their work in the 2000–2010 decade along diverse paths. Zalamea (2000–2006) profited from continuity and triadicity to propose novel perspectives on Latin American cultural studies. Niño (2008) wrote an extended chronological and critical Ph.D. Thesis on the development of abduction, without doubt the deepest study available on the theme at an international level. Andrade (2007–2009) advanced in his construction of a theory of biosemiotics based on Peirce's categories. And, above all, Oostra founded in 2007 his *Seminario Permanente Peirce* at the Universidad del Tolima (22 Thesis on Peirce), where his school on Peirce's mathematical logic has flourished (binary connectives, triadic logic, diagrams, intuitionism, existential graphs), to become the leading World center specialized in Peirce's graphs.

Time was ripe to organize the community and Zalamea created the *Centro de Sistemática Peirceana* (CSP) in 2007. The CSP has oriented its main task to the production of a yearly journal devoted to Peirce, the *Cuadernos de Sistemática Peirceana*. The first five numbers of the journal were written in Spanish by the local Colombian community of the CSP: Eugenio Andrade (biology), Gonzalo Baquero (philosophy), Carlos Garzón (philosophy), Lorena Ham (linguistics), Richard Kalil (philosophy), Jaime Lozano (economy), Alejandro Martín (mathematics), Douglas Niño (semiotics), Arnold Oostra (mathematics), Roberto Perry† (phonetics), Laura Pinilla (medicine), Miguel Ángel Riaño (philosophy), Edison Torres (philosophy), Fernando Zalamea (mathematics). One can consider a small feat the organization of such a multiverse community and its capacity to maintain a journal devoted specifically to Peirce (a unique fact, since even the *Transactions* extends itself towards general American philosophy). In contraposition with the initial local perspective, the next three numbers of the *Cuadernos* were oriented to articles by the global community (many languages represented, not just English) around monographic numbers: Esthetics (2014), Mathematics (2015), Existential Graphs (2016). Peirce would certainly have been intrigued to see a devoted community working on his heritage in a remote country that he never would have dreamed of.

This Volume

This volume concentrates on the *Mathematics of Peirce*, technically studied by the *Colombian School*. The three main topics concerned are the mathematizations and proofs of Peirce's **Pragmaticist Maxim** (Gustavo Arengas, *Chapter 1*), the construction and development of a full model for Peirce's **Continuum** (Francisco Vargas, *Chapter 2*), and the formal intuitionistic and geometrical extensions of Peirce's **Existential Graphs** (Arnold Oostra, *Chapter 3*). All these contributions are *firsts*, and entirely original. Some of them should come as gigantic surprises for the community of Peirce scholars. A short assessment of these achievements is offered as a final contribution (Fernando Zalamea, *Chapter 4*).

Chapter 1 offers the first formal axiomatization of *Peirce's Pragmaticist Maxim*, using sophisticated ideas, arguments, and variations of the machinery of Category Theory. The use of relational tools comes as a natural technique to capture the Maxim, but Arengas sharpens much the strata and layers of our comprehension of Peirce, thanks to his systematic use of refined categorical coverings (adjunctions, monads, descent theory). Not only the multiplicity of the Pragmaticist Maxim *explodes* in several formal contexts, but it becomes susceptible of *proof* in all of them, something that Arengas carries out with great technical precision. In this way, a strong *Mathematical Metaphysics* is obtained, something which would have bewildered the very Peirce.

Chapter 2 presents the first full modelization of *Peirce's Continuum*, including in a single model the generic, reflexive, supermultitudinous, and modal characters of the continuum. Vargas's construction uses the full range of ordinals in ZFC, and inverts their welding, in order to ensure the reflexivity of the model. Through an extremely simple ordinal iteration of the real line, and, subsequently, its inversion, precise definitions are provided, and theorems are proved. As a consequence of the construction of the model, ordinal layers of infinitesimals appear naturally in the model, helping to construct a very powerful differential and integral calculus. Further, the iterated layers can be seen as fragments of a modal Kripke model, and can address some mereological issues. All in all, Vargas's full model for Peirce's continuum puts at rest many philosophical speculations, and *mathematically proves*, once and for all, that Peirce's apparently wild ideas on the continuum are perfectly consistent.

Chapter 3 develops with great care the first formal proofs for *Intuitionistic Existential Graphs*, and explores the first extensions of the graphs to *Non-Planar Surfaces* (sphere, torus, cylinder, Möbius strip). The recognition that intuitionistic logic was the *natural* environment to understand Peirce's graphs, being both particular instances of a general logic of continuity, where logical and topological perspectives coincided, fostered Oostra's fastly growing research on the graphs. Three methods (diagrammatization, algebraization, linearization) were systematically superposed to obtain a full range of intermediate completeness theorems for superintuitionistic (and modal) logics. Further, the pendulum forces between the nice local character of surface logics, and their obstructive global behavior (with many negations/cuts, non classically interpretable), oriented Oostra's studies in the graph extensions to non-planar surfaces. Both the intuitionistic and geometric perspectives explore the multivalent meaning of deep connections between the logical and the topological that drive the *Mathematics of Peirce*.

Chapter 4 synthesizes the main ideas of the contributions by Arengas, Vargas, and Oostra. Profiting also from some beautiful and simple diagrams by Angie Hugueth, which capture some of the central forces exhibited in Arengas, Vargas, and Oostra's *gestures*, we present a compact view of the book along a conceptual *Riemann Surface*. The four sheets of the surface for $\sqrt[4]{z}$ integrate five generations: Zalamea (b. 1959), and four descending layers of his students, Oostra (b. 1966), Vargas (b. 1977), Arengas (b. 1989), Hugueth (b. 1998). As a result, a synthetic view of *Advances in Peircean Mathematics: The Colombian School* is provided.

References

The Bibliography is divided into two parts: *Peirce Bibliography* and *Secondary Bibliography*. References to Peirce's works are given in the form [**years**], invoking publication dates of the articles or compilations, followed by mentions to subsequent pages or paragraphs; each entry in the Bibliography has appended a list of all the pages where the reference is used throughout the book. The usual acronyms (CP, EP, NEM, etc.) are added in the citations to allow a quick glance for the scholars; for example, [**1931–58**] CP 5.402 refers to *Collected Papers*, published between 1931 and 1958, volume 5, paragraph 402. The *Secondary Bibliography* is treated as usual; all references are systematically provided in the footnotes.

Fernando Zalamea, Bogotá, November 2021.

Gustavo Arengas

1 Category Theory Variations and Proofs of Peirce's Pragmaticist Maxim

Abstract: We study pragmatism and category theory as (partial) resolutions of a transit/obstruction dialectic. We show how mathematical counterparts of Peircean thought (such as the triadic character of the sign, the pragmatic maxim, the modalities) arise naturally, and how formal theorematics can shed light on some parts of the pragmaticist architecture.

Keywords: Lautman; category theory; sign; pragmatic maxim; modalities

Peirce's Pragmaticist Maxim forces immediately on the Reader an acknowledgment of an infinite hierarchy of layers, perspectives, and interpretations. To be able to render, study, define, and prove some statements about that hierarchy, it becomes natural to use category-theoretic tools in a formalization of the problems involved. *Chapter 1* addresses part of those issues. *Section 1.1* surveys some correspondences between Category Theory, Physics, Topology, Logic, and Computation, which help to understand the Pragmaticist Maxim in a holistic way. *Section 1.2* introduces, and develops, the category-theoretic tools (simplices, higher categories, ∞-categories) needed to formalize the sign processes involved in the Maxim. *Section 1.3* offers a dynamic interpretation of the Maxim in functorial terms. *Section 1.4* profits from advanced category-theoretic techniques (monads and comonads) to define categorically habits and effects, and further formalize the Maxim. *Section 1.5* combines sheaves and monads to analyze the Maxim (descent theory) into its many adjointness and reflection levels. Finally, *Section 1.6* studies the progressive determination levels of physical reality, related to the many adjunctions already present in the Maxim.

Acknowledgment: To Fernando Zalamea for his support, patience and perseverance to carry out this project. To Arnold Oostra and Francisco Vargas for their inspiring work on which this chapter draws so much. To Angie Hugueth for her beautiful graphic synthesis that captures very well the ideas raised here.

Gustavo Arengas, Gustavo Arengas (1989) has been Instructor at Universidad Nacional de Colombia, and is now Professor at Universidad Popular del Cesar.

https://doi.org/10.1515/9783110717631-001

1.1 Unifying Correspondences

From the origins of computer theory and intuitionistic mathematics in the first decades of the 20th century, similarities between writing a program and making a constructivist proof began to be observed. Over the years, when the structures in question were developed, these analogies became a precise theorematic that relates certain constructions that have arisen in computational languages and mathematical logic. With the development of category theory, these analogies were consolidated and refined thanks to the new language. For example, we have that given a theory $\mathbb{T}$ in a first order language, we can form a category $\mathcal{C}_\mathbb{T}$, called the syntactic category of $\mathbb{T}$. All sentences of $\mathbb{T}$ can be interpreted and verified in $\mathcal{C}_\mathbb{T}$ (Soundness Theorem) and all the propositions that are valid in the same type of categories as $\mathcal{C}_\mathbb{T}$ are in $\mathbb{T}$ (Completeness Theorem). This construction allows us to convert logical notions into categorical notions (and vice versa): true/terminal object, false/initial object, conjunction/product, implication/internal hom, etc. This class of transits is known as Curry–Howard–Lambek (CHL) correspondence, or computational trinitarianism.

There have been several proposals to extend this relationship to other fields. Baez and Stay[1] add quantum mechanics and topology from a good notion of a tensor product ⊗. They call the new relationships *The Rosetta Stone*, in remembrance of the stela written in hieroglyphics, Demotic and Greek, which was essential to understanding the language of Ancient Egypt. To give the reader an idea, the Table 1.1 shows the correspondence proposed.

Table 1.1: The Rosetta Stone

	Category Theory	Physics	Topology	Logic	Computation
X	Object	Hilbert space	Manifold	Proposition	Data type
$f: X \to Y$	Morphism	Operator	Cobordism	Proof	Program
$X \otimes Y$	Tensor product of objects	Hilbert space of joint system	Disjoint union of manifolds	Conjunction of propositions	Product of data types
$f \otimes g$	Tensor product of morphisms	Parallel processes	Disjoint union of cobordisms	Proofs carried out in parallel	Programs executing in parallel

---

[1] Baez and Stay 2011.

The authors hoped that just as the former Rosetta Stone allowed translations between the different languages of ancient Egypt, the new one would allow us to move ideas and constructions between contexts where we had a good notion of process; for example, categorical notions such as arrow composition and tensor product correspond to the idea of serial processes and parallel processes.

More recently Gorard et.al.,[2] have re-studied the correspondence, but this time motivated by a possible reunification of the laws of physics through hypergraphs that are transformed by rewriting rules that obey the Church–Rosser property, known as Wolfram models. According to this theory, all the fundamental concepts of physics (space, time, matter, energy, etc.), can be understood as properties of the hypergraph. In particular, from the ambiguity of rewriting laws on the hypergraph, where lines of evolution bifurcate, arises our quantum-relativistic apprehension of the universe.

Table 1.2: Correspondence in Gorard

Logic	Physics	Computation
Axioms	Initial state	Input
Proof	Motion	Computation
Rules of inference	Laws of motion	Algorithmic rules
Theorem	Final state	Output

As can be seen, what is emerging is, *a general science of systems and processes*[3] which focuses on the notion of transition and the mechanisms to identify it. In the previous approximations, the tensor product and the higher-dimensional arrows help us to glimpse the growing and dizzying network of divergent, convergent and parallel transits. In this same line, in his distillation of pragmaticism in the synthetic, Maddalena[4] shows how the main characteristic of Peircean thought is the study of tools to recognize the identity of the interpretative processes through change. These tools are everywhere in Peirce's work. First, around the concept of sign: as we will discuss in more detail below, a sign can be thought of as a process that communicates forms of its object to its interpreter; also, in the formation of the ultimate intellectual interpreter (habit), as the limit of a chained semiosis; or in the process determination of the vague/general to the determined; or in the

---

2 Wolfram 2020.
3 Baez and Stay 2011, p. 97.
4 Maddalena 2015.

evolution of existential graphs within the assertion sheet. This last example plays a central role within the pragmaticist architecture, especially since Peirce raised the possibility that the graphs could provide a proof of pragmatism. Let us review what the latter would consist of:

> You apprehend in what way the system of Existential Graphs is to furnish a test of the truth or falsity of Pragmaticism. Namely, a sufficient study of the Graphs should show what nature is truly common to all significations of concepts; whereupon a comparison will show whether that nature be or be not the very ilk that Pragmaticism (by the definition of it) avers that it is. It is true that the two terms of this comparison, while in substance identical, yet might make their appearance under such different garbs that the student might fail to recognize their identity [...] But, on the other hand, should the theory of Pragmaticism be erroneous, the student would only have to compare concept after concept, each one, first, in the light of Existential Graphs, and then as Pragmaticism would interpret it, and it could not but be that before long he would come upon a concept whose analyses from these two widely separated points of view unmistakably conflicted [...][5]

The above quote states that pragmatism is true if and only it is in good correspondence with the existential graphs. Already Brady and Trimble[6] have shown the close relationship between existential graphs (at least the *Alpha* and *Beta* part) with monoidal categories and string diagrams, which is precisely the framework of the *Rosetta Stone* following Baez and Stay. To fully understand this situation, it is useful to take up some approaches of the French mathematical philosopher Albert Lautman, who, in the 1930s, proposed a re-understanding of mathematics based on a dynamic Platonic dialectic. According to Lautman: "The reason for the relations between Dialectics and Mathematics resides in the fact that the problems of Dialectics are conceivable and formulable independently of Mathematics, but, at the same time, in that every sketch of solution contributed to these problems is necessarily based on some mathematical example, destined to support concretely the dialectical link studied".[7] Using the terminology of this French philosopher who defines "*notion* as one of the poles of a conceptual tension, and an *idea* as a partial resolution of this polarity",[8] Fernando Zalamea's work shows how the ideas associated with certain fundamental dialectics (*transit/obstruction, local/global, multiple/one*, among others) are not only present in contemporary

---

5 [1931–58] CP 4.534, note 1.
6 Brady and Trimble 2000, Brady and Trimble n.d.
7 Lautman 2011, p. 379.
8 Lautman 2011 p. 49, p. 334.

mathematics,[9] but also in Peircean pragmatism[10] and, in general, in the richest fruits of the human spirit.[11]

Probably without realizing this philosophical background, the various versions of the Curry–Howard correspondence (Tables 1.1, 1.2) show how, when faced with similar dialectics, the different sciences have given rise to similar ideas. In the same way, the proof of pragmatism claimed by Peirce in the previous quote, would show how the ideas expressed in the entire architecture are perfectly reflected (in the sense of the Peircean continuum, where *any part of which however small itself has parts of the same kind*[12]) in existential graphs. According to Lautman, the task of the philosophy of science should be to build a Theory of Ideas, which includes, among other investigations, the description of the ideal structures embodied in the different disciplines, to establish a hierarchy and a genesis of ideas, some arising from others, and to show, within the ideas themselves, the reasons for their application to the sensible universe.[13] In this work we limit ourselves to show the good correlation between some ideas of pragmatism and category theory, with sporadic mention of other branches of correspondence in some key points.

As a mouth opener, let's review one of the most revealing explanations given by Peirce on the concept of sign:

> [...] a Sign may be defined as a Medium for the communication of a Form [...] As a medium, the Sign is essentially in a triadic relation, to its Object which determines it, and to its Interpretant which it determines [...] That which is communicated from the Object through the Sign to the Interpretant is a Form. It is not a singular thing; for if a singular thing were first in the Object and afterward in the Interpretant outside the Object, it must thereby cease to be in the Object. The Form that is communicated does not necessarily cease to be in one thing when it comes to be in a different thing, because its being is a being of the predicate. The Being of a Form consists in the truth of a conditional proposition. Under given circumstances, something would be true. The Form is in the Object, entitatively we may say, meaning that that conditional relation, or following of consequent upon reason, which constitutes the Form, is literally true of the Object. In the Sign the Form may or may not be embodied entitatively, but it must be embodied representatively, that is, in respect to the Form communicated, the Sign produces upon the Interpretant an effect similar to that which the Object itself would under favorable circumstances.[14]

---

9 For instance, Zalamea 2012a, and Zalamea 2019.
10 For instance, Zalamea 2010c, chapter 1, and Zalamea 2014, chapter 5.
11 For example, in Zalamea 2010c, Zalamea studies Novalis, Beethoven, Turner, Valéry, Proust, Kiefer.
12 [1982–2009] W 3.103, "The Conception of Time Essential in Logic" (1873); Zalamea 2012a, p. 16.
13 Lautman 2011, pp. 382–383.
14 [1999–98] EP 2.544.

If we denote the object with O and the interpretant with I, the previous quote motivates us to think of a sign as a function, or in more categorical terms, as a morphism $S : O \to I$, which communicates (or transports) *forms* from O to I. However, it is revealing that Peirce refers to a sign as a medium that allows communication, because for this to be possible the medium must necessarily be continuous: if there were a gap in it, all communication would be interrupted. Given the role that continuity plays in his thinking, this sense should not have gone unnoticed by Peirce when choosing his words. In *Section 1.2* we will see how, by applying hypostatic abstraction to the triadic character of the sign, it can be shown that the medium between the object and the interpretant must be effectively continuous. For this we will use a series of works by many mathematicians (including Quillen, Grothendieck, Joyal) who have revealed the close relationship between topological structures and higher-dimensional categories.

However, such a medium, which has in itself the continuity that allows transit, has also a certain viscosity, an embodiment of obstruction, which interferes with the transfer of information from the object to the interpretant, even causing it to fail on some occasions. From this resistance to transit arises the modal character so characteristic of Peircean thought, impregnated everywhere with *will be*, *would be*, etc. This seriously ruins our approach of treating a sign as a certain function $S : O \to I$, since such a function must always carry every value from O to I. The same problem is observed in the theory of computation: as the Curry–Howard correspondence shows (Tables 1.1 and 1.2), a program can be thought of as a function except that, as anyone who has ever used a computer has observed, many times it fails to return an output. A rather elegant solution to this problem was given by the Italian computational scientist Eugenio Moggi,[15] who proposed to model these impurities using monads. The use of monads not only allows us to model the obstructions of the environment to the transit of information, but also shows how the same dialectic acts in other sectors of Peircean thought, such as in the pragmatic maxim, the formation of habits, or the process of determination. The emergence of the same scheme in such different fields of knowledge can only be explained, as Lautman observed, from a much deeper dialectic that humanity has been revealing when trying to apprehend the transitory character of the universe.

A fundamental aspect to highlight about this approach is that the ideas that appear in this correspondence are completely *natural*, that is, forced by the context where they emerge. Perhaps, as Peirce students, we have only *failed to recognize their identity*. We will see how all the categorical constructions that we will

---

[15] Moggi 1989.

use have their germs in the natural original ideas of the theory (decades of the 1950s and 1960s).

1.2 Signs and ∞-Categories

As mentioned above, the Peircean definition of sign as a Medium for the communication of a Form of the object to the interpreter, motivates us to represent it by a morphism $S : O \to I$, with O, I denoting the object and the interpretant of the sign, respectively. Now, resorting to hypostatic abstraction, "that process whereby we regard a thought as a thing", Peirce argues that we can consider each sign as the object of a new sign.[16] This fact has an important *geometric consequence* for a possible category of signs that we would like to consider: on it we must be able to define objects of ever larger *dimensions*. To see this, suppose that S_0 is an arbitrary sign, which can be considered as an object of another sign $S_1 : S_0 \to S_0'$. The key point is that in this situation S_1 must have one dimension more than S_0. If we imagine, using our geometric intuition, S_0 as a point, S_1 would be a one-dimensional arrow. Similarly, if we consider S_1 as the object of a sign $S_2 : S_1 \to S_1'$, the latter will have one dimension more than S_1, thus being two-dimensional. A S_3 that has S_2 as an object must be three-dimensional and so on, *ad infinitum*. Therefore, beyond traditional category theory, we need a more sophisticated version, known as *higher category theory*, which allows us to work with multidimensional entities. Many different candidates have been proposed to formalize the intuitive idea of what such a theory should be; here, as it is the most developed and most widespread of the existing ones, we will only work with the model based on simplicial sets, known as *quasi-categories*.

Definition 1.1. The simplicial category Δ is given by
- objects: non-empty totally ordered sets $[n] := \{0 \le 1 \le \ldots \le n\}$ for $n \ge 0$;
- morphisms: order-preserving maps. A more convenient notation is to write $\langle m_0 \ldots m_n \rangle : [n] \to [m]$ for the function $i \mapsto m_i$.

The category of simplicial sets is the category of presheaves on Δ, *i.e.*, functors $\mathcal{S} : \Delta^{op} \to Sets$.

Given any simplicial set $\mathcal{S}$, we will write $\mathcal{S}_n := \mathcal{S}([n])$ and call the elements of the set $Ob(\mathcal{S}) = \coprod_{n \in \mathbb{N}} \mathcal{S}_n$ the *objects* of $\mathcal{S}$. These objects can be classified as non-degenerate or degenerate. As their names suggest, non-degenerate objects

---

16 [1931–58] CP 5.448, 4.549.

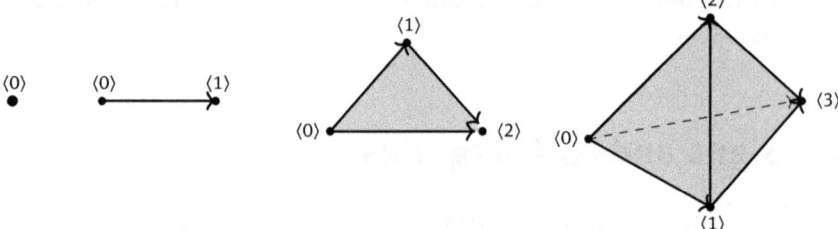

Figure 1.1: First simplices

generate degenerate ones, so when we graph a simplicial set it is enough to focus only on the former. For the representable functors $\Delta^n := Hom_\Delta(\_, [n])$, called *simplices*, the non-degenerate objects are the strictly monotonic functions. In the following image (*Figure 1.1*), we draw this type of objects for the simplices Δ^0, Δ^1, Δ^2, Δ^3. For example, the non-degenerate objects of Δ^2 are, using the notation from Definition 1.1, the functions $\langle 0 \rangle$, $\langle 1 \rangle$, $\langle 2 \rangle : [0] \to [2]$ of one-dimensional range, $\langle 01 \rangle$, $\langle 12 \rangle$, $\langle 02 \rangle : [1] \to [2]$ of two-dimensional range and $\langle 012 \rangle : [2] \to [2]$ of three-dimensional range, which we represent graphically by the three points, the three arrows and the gray triangle in the third image. Since for any simplicial set $\mathcal{S}$, Yoneda's Lemma establishes natural bijections $\mathcal{S}_n \simeq Hom_{sSet}(\Delta^n, \mathcal{S})$, we can think the objects of $\mathcal{S}_0$ as points, those of $\mathcal{S}_1$ as arrows, those of $\mathcal{S}_2$ as triangles, etc. In this way, as the notation suggests, the subscripts effectively indicate the dimension of the different objects in the simplicial set. The Δ morphisms allow transits between objects of different dimensions. For example, we have two functions $\mathcal{S}_{\langle 0 \rangle}$, $\mathcal{S}_{\langle 1 \rangle} : \mathcal{S}_1 \to \mathcal{S}_0$ such that $\mathcal{S}_{\langle 0 \rangle}(x \to y) = x$, $\mathcal{S}_{\langle 1 \rangle}(x \to y) = y$. Similarly, we have $\mathcal{S}_{\langle 01 \rangle} : \mathcal{S}_0 \to \mathcal{S}_1$ such that $\mathcal{S}_{\langle 01 \rangle}(x) = id_x : x \to x$.

One of the greatest assets of Peircean semiosis is that it allows the concatenation of signs: the interpreter of a sign can be seen as the object of another sign. In our analogy of representing a sign by an arrow, this corresponds precisely to the composite of maps. Given the importance of this operation in the traditional theory, it is expected that it can be generalized to higher category theory. However, not all simplicial sets have an adequate composition operation. To characterize those that do, we introduce the horn Λ^n_i, $0 \le i \le n$, which is the subfunctor of Δ^n given by $(\Lambda^n_i)_k = \{f : [k] \to [n] \mid [n] - \{i\} \not\subseteq f([k])\}$. Intuitively, the nondegenerate objects of Λ^n_i are obtained by eliminating the interior of Δ^n and the opposite face of i. Thus, the three horns of Δ^2 can be represented graphically as in *Figure 1.2*. A simplicial set will be a ∞-category (or a *quasi-category* in Joyal's original terminology) if the image of certain horns can be completed.

1 Category Theory Variations and Proofs of Peirce's Pragmaticist Maxim — 9

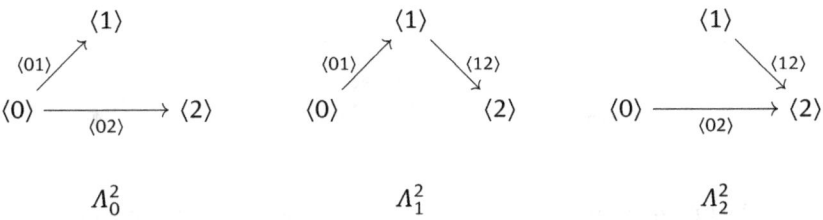

Figure 1.2: Horns of Δ^2

Definition 1.2. A ∞-category is a simplicial set S which has the following property: for any $0 < i < n$, any map $\alpha' : \Lambda_i^n \to S$ admits an extension $\alpha : \Delta^n \to S$.

These ∞-categories are the higher dimensional analogs of traditional categories. There are several aspects of the above definition that we would like to highlight. First, let's note that this extension property does capture the composition operation of traditional category theory. To see this, let's consider a ∞-category S and two morphisms $f : x \to y$, $g : y \to z$ in S_1. To define a function $\alpha' : \Lambda_1^2 \to S$ just send the non-degenerate objects $\langle 0 \rangle$, $\langle 1 \rangle$, $\langle 2 \rangle$, $\langle 01 \rangle$, $\langle 12 \rangle$ (represented in Figure 1.2) to x, y, z, f, g, respectively. The existence of *an* extension $\alpha : \Delta^2 \to S$ of α' is equivalent, thanks to Yoneda's Lemma, to the existence of an object in S_2, which, following the usual convention in category theory, we will continue to denote α. Thanks to the naturality of the bijection, when we project this α properly we get x, y, z, f, g. Also, the projection $S_{\langle 02 \rangle} : S_2 \to S_1$ gives a new arrow $\alpha_{\langle 02 \rangle} : \alpha_{\langle 0 \rangle} \to \alpha_{\langle 2 \rangle}$, which, due to the fact that they all come from the same object α, we can consider *a* composition of f, g.

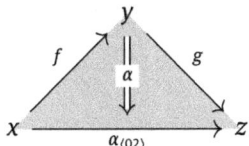

Notice that in Definition 1.2 we do not require the uniqueness of the extension, and therefore we cannot guarantee the uniqueness of the compound either. We can say that the assignment of a pair of maps to its composite is ambiguous. In fact, given a ∞-category S, we can associate it with an ordinary category hS, the homotopy category of S, defining that two morphisms $f, g : x \rightrightarrows y$ are equivalent if one can be *continuously deformed* into the other via some object of S_2:

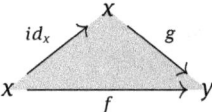

In the dialectic between S (with ambiguous, indeterminate composite) and hS (with definite, determinate composite) lies one of the main reasons for the theory: study homotopic structures directly. This dialectic, being also essential in pragmatism, largely explains the good correspondence between these two disciplines. For example, as Peirce had already observed, continuity in pragmatism arises from indeterminacy: *The continuum is a General. It is a General of a relation. Every General is a continuum vaguely defined.*[17] As we will see in this section, in higher category theory continuity also arises from the indeterminacy of the composite.

Next, we would like to look at the way ∞-categories have been introduced versus how traditional category theory is defined. In the latter, a category is defined from a set of axioms from which the theorems are deduced in the traditional way. On the other hand, in ∞-categories we only have multidimensional entities and operations that allow their transformations: extraction of faces, expanding them freely through a dimension and the possibility of filling boxes suitably assembled by Definition 1.2. Here the veracity of a proposition depends only on the possibility of constructing a diagram that attests it. For example, in traditional category theory, given a map $f : x \to y$ we have that $f \circ id_x = f$ thanks to one of the identity axioms, while in ∞-categories the proposition continues to be valid, but thanks to the fact that we can draw the gray area that proves it.

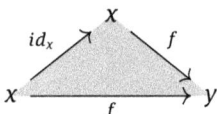

This relationship reminds us of Peirce's sentence, *The Being of a Form consists in the truth of a conditional proposition.*[18] If the forms testify to the truth of the propositions, then the operations that manipulate them become methods of reasoning. Thus, for example, the extension operation in Definition 1.2 corresponds to an *abduction*:

---

17 [1976] NEM 3.925, "Letter to E. H. Moore" (1902).
18 [1999–98] EP 2.544, reproduced above on p. 5.

- This configuration of morphisms is horn-shaped.
- All the configurations of morphisms arising from simplices are horn-shaped.
- Therefore, this morphism configuration arises from some simplex.

Finally, let us highlight that in Definition 1.2 we do not require the extension to be for all i, but only for those that are inside the discrete interval $[0, n]$. Those where the property is valid without restriction have very important properties, as we will see next.

Definition 1.3. A ∞-category $\mathcal{S}$ is called a ∞-groupoid if in Definition 1.2 the extension property also applies when i is equal to n or when it is equal to 0 (both conditions are equivalent).

It is helpful to put the terminology in context. Given a group G (a set with an associative operation, with identity and inverses) we can define a category with a single object, say $\star$, such that $Mor(\star, \star)$ is by definition equal to G. These categories have the following property that characterizes them: every map is an isomorphism. In analogy to this situation we say that a category (no longer necessarily with a single object) is a groupoid if all its morphisms are isomorphisms. The fact that ∞-groupoids constitute the proper categorical higher version of groupoids is given by

Proposition 1.1. A simplicial set $\mathcal{S}$ is a ∞-groupoid iff $\mathcal{S}$ is a ∞-category and its homotopy category $h\mathcal{S}$ is a groupoid.

To fully prove this fact would require discussing the theory a little more than what we propose in this work. However we can show here that in a ∞-groupoid $\mathcal{S}$ all morphisms have an inverse (not necessarily unique). First, we consider any map $(f : x \to y) \in \mathcal{S}_1$ and define the horn function Λ_0^2 to $\mathcal{S}$, which sends the side $\langle 01 \rangle$ to f and $\langle 02 \rangle$ to id_x. This can be extended to Δ^2, attesting to the existence of a $g : y \to x$ such that $gf \simeq id_x$.

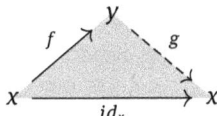

To see that $fg \simeq id_y$, we form a function of Λ_0^3 with image in $\mathcal{S}$ given by

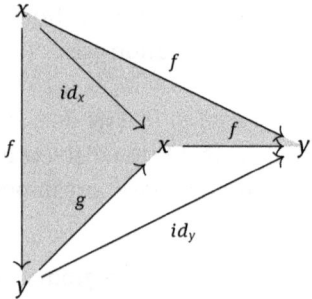

This can be extended to Δ^3, which attests that the white triangle in the diagram above also commutes.

Given their introduction as a generalization of the concept of group, the groupoids appear in a great variety of situations. Here we are especially interested in its close relationship with topological spaces. The idea dates back to the French mathematician Henri Poincaré, who at the end of the 19th century presented two mechanisms to study spaces algebraically: homology and homotopy. In the latter, we try to characterize a space X through the behavior of the paths that we can define on it. A path on X is a function on the unit real interval $p : [0, 1] \to X$, the direction of which is determined by the order of $[0, 1]$. Two such paths p, q can be identified if there is a homotopy between them, that is, if we can continuously deform one into the other within X. The path equivalence classes not only constitute a category (with the points of space as objects and the classes as maps), but also a groupoid (with the inverse of a morphism obtained by reversing the order of $[0, 1]$ of any of the members of the class). The fundamental question of algebraic topology is how characteristic this algebraic information is with respect to the original topological space. In the language of category theory, this corresponds to asking how close are the categories $\mathcal{TOP}$ (of topological spaces and continuous functions) and $\mathcal{GPD}$ (of groupoids and functors), and, in particular, if they are equivalent. When working with traditional groupoids, the equivalence only applies to a fairly restricted subcategory of $\mathcal{TOP}$. However, when moving to higher dimensions the situation improves considerably and we obtain:

Proposition 1.2. There is an equivalence of ∞-categories between $\infty - \mathcal{TOP}$ and $\infty - \mathcal{GPD}$:

1 Category Theory Variations and Proofs of Peirce's Pragmaticist Maxim — 13

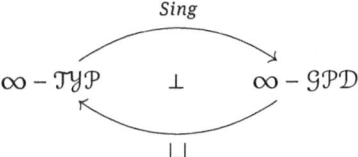

where $\infty - \mathcal{TYP}$ is the ∞-category of homotopy types, with: CW-complexes as 0-objects, continuous maps as 1-objects, homotopies as 2-objects, homotopies between homotopies as 3-objects, homotopies between homotopies between homotopies as 4-objects, ...

A CW-complex corresponds to what we could think as a nice topological space, built by assembling discs of various dimensions. The equivalence can be made explicit through topological analogs of the Δ^n, seen no longer as representable functors, but as subspaces of $\mathbb{R}^{n+1}$. For this we define the functor $\Delta_{Top} : \Delta \to \infty - \mathcal{TYP}$, which sends each $[n]$ to $\Delta_{Top}^n = \{(x_0, \ldots, x_n) | \sum_{i=0}^n x_i = 1, \; x_i \geq 0\}$. From here, to any space X we assign the ∞-groupoid $SingX : \Delta \to \mathcal{SET}$, $[k] \mapsto Hom_{Top}(\Delta_{Top}^k, X)$, and to any ∞-groupoid S its geometric realization $|S|$, which is constructed by interpreting each object in S_n as a copy of Δ_{Top}^n and pasting all of them by their borders using the information from the morphisms of Δ (face and degeneracy maps).

Thus, each ∞-groupoid can be thought of as a nice topological space. Here another great difference is obtained in the higher category theory realm with respect to the traditional one: while in the second the *discrete*, represented by the category of sets, plays a primordial role in many categorical constructions, in the first one this role is transferred to the *continuous*, that is, to $\infty - \mathcal{GPD}$. For example, given two 0-objects x, y in a ∞-category S, the morphisms from x to y, $Mor(x, y) = \{f \in S_1 | S_{(0)}(f) = x, \; S_{(1)}(f) = y\}$ form an ∞-groupoid. In effect, S induces on $Mor(x, y)$ several equivalent higher structures. One of them, the right morphisms, consists in considering as $z \in Mor(x, y)_n$ those $z \in S_{n+1}$ such that $S_{(n+1)}(z) = y$ and $S_{(0^n)}(z) = x$ (where, $\langle 0^n \rangle : [n] \to [n+1]$ is the function constant at 0). Thus, an arrow from f to g is a $\theta \in S_2$ of the form:

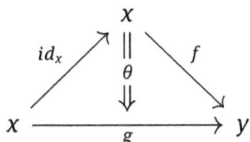

Any of the 1-objects of *Mor(x, y)* has an inverse: given $\theta : f \to g$, we form the map of the horn Λ_2^3 that sends the sides $\langle 13 \rangle$, $\langle 23 \rangle$ to g, f. The face in front of $\langle 2 \rangle$ in the image of the lifting of the map to Δ^3 provides the inverse $\theta^{-1} : g \to f$:

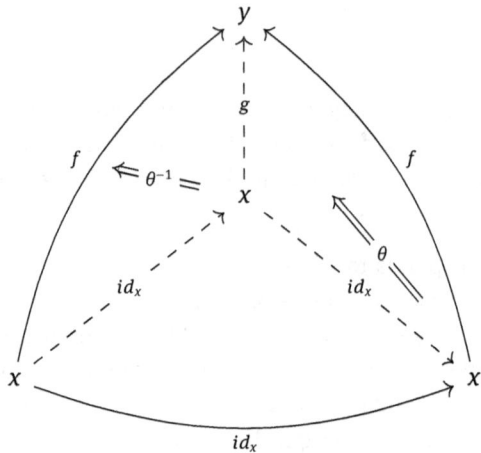

Thus, in higher categories *Mor(x, y)* is naturally provided with a notion of continuity: intuitively the existence of a morphism $\theta : f \to g$ means that we can continuously deform f to g. The richness of this topology depends, as we mentioned earlier, on the ambiguity of the compound: if it were univocally defined, the only possible θ would be the identity and the resulting topology would be the discrete one; while if there were much more information at the higher levels, *whose possibilities of determination no multitude of individuals can exhaust*,[19] we would be in the presence of a true continuum.

As we will discuss in detail in the following sections, information in Peircean semiotics does not travel in a single track from the object to the interpreter, but rather is a pendulum process, with multiple oscillations, where in each of them we collect new knowledge. This transit is only possible because the space between them is continuous (actually, as we will see in the last section, it is rather a continuum under construction, which is enriched with each new oscillation).

---

[19] [1931–58] CP 6.170.

1.3 Unlimited Semiosis and (co)Limits

Let us focus for now on the interpretant of the sign, the effect caused by the sign on its meaningful capacity. Studying these effects in detail in his theory of the *phaneron*, "the collective total of all that is in any way or in any sense present to the mind, quite regardless of whether it corresponds to any real thing or not",[20] Peirce distinguishes three classes of interpreters in correspondence to their three cenopythagorean categories[21]:

1. *Feeling*. Peirce gives the following example: "an air for a guitar, if considered as meant to convey the genuine or feigned musical emotions of its composer, can only fulfill this function by exciting responsive feelings in the listener. But in the second place, the interpretant may be an effort."[22]
2. *Effort*. "Thus, when a drill-officer gives a company of infantry the word of command, "Ground arms!", if this is really to act as a sign and not in a purely "physiological" way (I use this inaccurate distinction, rather than waste time in explanations), there must first be, as in all action of signs, a feeling-interpretant,—a sense of apprehending the meaning,—which in its turn at once stimulates the soldiers to the slight effort required to perform the motion. This effect caused by the sign in its significative capacity is, by the definition, an interpretant of it."[23]
3. *Triadic interpretant*. It is "the interpretant *par excellence*, [...] the intellectual apprehension of the meaning of a sign"[24].

The examination of this triadic interpretant constitutes, according to Peirce, the key to pragmatism. Resorting again to hypostatic abstraction, it is shown that the interpretant of any sign is again a sign, which has an interpretant that is in turn

---

20 [1931–58] CP 1.284.
21 Following Zalamea 2012a, p. 58, Peircean categories can be described with key words and fundamental concepts as follows:
1. *Firstness*: immediacy, first impression, freshness, sensation, unary predicate, monad, chance, possibility.
2. *Secondness*: action–reaction, effect, resistance, alterity, binary relation, dyad, fact, actuality.
3. *Thirdness*: mediation, order, law, continuity, knowledge, ternary relation, triad, generality, necessity.

22 [1999–98] EP 2.430.
23 [1999–98] EP 2.430.
24 [1999–98] EP 2.430.

a new sign, which gives rise to a new interpretant, in an *ad infinitum* process.[25] Peirce thought that this process always converged, and tended to the ultimate intellectual interpreter, which by its very construction as an "accumulation of interpreters", corresponds to habit. We will return to this point later. From the argument we have just seen, it is clear that this type of interpretant should be modeled by a notion of a higher dimensional (co)limit (by viewing each interpretant as a new sign, we change the dimension of the entity).

There are several ways to introduce (co)limits in traditional category theory. A rather elegant one, which has the advantage of being able to be directly generalized to higher dimensions, makes use of the join ($\star$) operation, which for two categories A, B is defined as:

$$Ob(A \star B) = Ob(A) \sqcup Ob(B)$$
$$Mor(A \star B) = Mor(A) \sqcup (Ob(A) \times Ob(B)) \sqcup Mor(B).$$

Thus, $A \star B$ is simply the disjoint union of the two categories, plus unique morphism of all the objects of the first into those of the second. Let us denote with s_A, s_B the respective inclusions of A, B in $A \star B$. Given a functor $F : I \to S$, we define the slice category $S_{F/}$:
- objects: functors $G : I \star \Delta^0 \to S$ such that $G|_A = F$
- morphisms of G to G': functors $G : I \star \Delta^1$ such that $H|_{I \star \langle 0 \rangle} = G$, $H|_{I \star \langle 1 \rangle} = G'$,

where, thanks to its graphic representation (*Figure 1.1*), we denote by Δ^0 the category with a unique object and by Δ^1 the category with two objects and a single map between them. By this construction, a cocone is an object of $S_{F/}$, a limit cocone an initial object $\widehat{G}$ of $S_{F/}$ and the colimit is $\widehat{G}|_{\Delta^0}$.

To generalize this construction to higher category theory, it is enough to consider larger dimensions. Thus, given two simplicial sets A, B their joint is defined as

$$(A \star B)_n := \bigsqcup_{[n]=[n_1]\sqcup[n_2]} A_{n_1} \times B_{n_2}$$

where $[p] \sqcup [q] := [p + q + 1]$. Then

$$(A \star B)_0 = A_0 \sqcup B_0$$
$$(A \star B)_1 = A_1 \sqcup (A_0 \times B_0) \sqcup B_1$$
$$(A \star B)_2 = A_2 \sqcup (A_1 \times B_0) \sqcup (A_0 \times B_1) \sqcup B_2$$
$$\vdots$$

---

25 [1999–98] EP 2.430.

Thus, again $A \star B$ consists of combining the n-objects of A with those of B and projecting the former into the latter. Given a simplicial set map $F : I \to S$, G will be an n-dimensional object of the slice category $S_{F/}$ iff the following diagram commutes

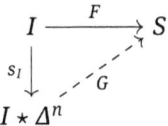

With these extensions, it is enough to define, in the same way as in the traditional case, a cocone as an object of $S_{F/}$, a limit cocone as an initial object $\widehat{G}$ of $S_{F/}$ and the colimit as $\widehat{G}|_{\Delta^0}$.

Let us reread the previous construction in pragmatic terms. Suppose that S corresponds to the (∞)-category of all signs, while I and F indicate the relationship "be an interpreter of" in the semiotic process that we are considering. So, for example, the image of $k : i \to j \in Mor(I)$ in S tells us that "Fj is an interpreter of Fi seen when we interpret it from the form Fk". Perhaps the simplest unlimited semiosis process we can imagine is a chain. In this case, we can consider $I = \mathbb{N}$ and thus obtain

$$F0 \longrightarrow F1 \longrightarrow F2 \longrightarrow F3 \longrightarrow \cdots \longrightarrow Fn \longrightarrow \cdots$$

The colimit in this case represents the end of the semiotic chain.

One of the main virtues of category theory is that it allows us to work with dual statements, and thus, in tune with the categorical spirit, we work together on apparently disparate situations. The dual of the colimits are the limits, which are obtained by reversing the order in the previous constructions.[26] In the Peircean context we are working on, let's see how the limit of the functor $F : I \to S$ corresponds precisely to the pragmatic maxim. Throughout his life, Peirce gave various versions of his maxim. As a guide for the discussion that we will do in the following sections, we reproduce some of them:

> [...] the rule for attaining the third grade of clearness of apprehension is as follows: what effects that might conceivably have practical bearings, we conceive the object of our con-

---

[26] We will not do it explicitly here for reasons of space, but the interested reader can consult Joyal 2002, Definition 4.5.

ception to have. Then our conception of those effects is the whole of our conception of the object.[27]

Pragmatism is the principle that every theoretical judgment expressible in a sentence in the indicative mood is a confused form of thought whose only meaning, if it has any, lies in its tendency to enforce a corresponding practical maxim expressible as a conditional sentence having its apodosis in the imperative mood.[28]

The entire intellectual purport of any symbol consists in the total of all general modes of rational conduct, which, conditionally upon all the possible different circumstances and desires, would ensue upon the acceptance of the symbol.[29]

(Pragmatism asserts), that the total meaning of the predication of an intellectual concept is contained in an affirmation that, under all conceivable circumstances of a given kind (or under this or that more or less indefinite part of the cases of their fulfillment, should the predication be modal) the subject of the predication would behave in a certain general way—that is, it would be true under given experiential circumstances (or under a more or less definitely stated proportion of them, taken as they would occur, that is in the same order of succession, in experience).[30]

One can immediately see from the above excerpts that the Pragmatic Maxim (PM) is linked to correlations, modalities, iterations, which in turn are prominent in Category Theory (CT). In what follows, we will make clear the connections between (PM) and (CT).

If $F : I \to S$ indicates the flow of a semiotic process, its limit l is an object of S such that:
- it is a cone, that is, it can be properly projected in the image of F. According to our correspondence, this means that all the Fi are interpretants of it, in such a way that the flow of the semiotic process given by F is respected.
- it is the limit cone, that is, l is completely determined by F. This corresponds precisely with the 1878 statement of the maxim, where it is established that any object is completely characterized by its effects, that is, the interpretants to which it gives rise.

Zalamea has suggested the close relationship between the pragmatic maxim and sheaf theory.[31] Let us see how, by refining our approach using Peircean ideas, these structures arise naturally. First, the interpretations of an object must consti-

---
27 [1931–58] CP 5.2.
28 [1931–58] CP 5.18.
29 [1931–58] CP 5.438.
30 [1931–58] CP 5.467.
31 For example in Zalamea 2012b, Chapters 3, 9.

tute a continuum.³² One of the most important properties of this continuum is its inextensibility, that is, it "cannot be composed of points":³³ in it "there is not room for them (the points) to retain their distinct identities; but they become welded into one another".³⁴ In a continuum we can only determine neighborhoods of points, each one of them, by reflexivity,³⁵ constituting in turn a continuum. It is in these neighborhoods where the interpretants flow:

> Logical analysis applied to mental phenomena shows that there is but one law of mind, namely, that ideas tend to spread continuously and to affect certain others which stand to them in a peculiar relation of affectibility. In this spreading they lose intensity, and especially the power of affecting others, but gain generality and become welded with other ideas.³⁶

Thus, our functor $F : I \to S$, rather than sending points to signs, should determine neighborhoods within the continuum flow of interpretants. Mathematically, this means that it constitutes an ∞-sheaf. Then, in the first place, I must no longer be an arbitrary category but must be made up of neighborhoods. The simplest way to do this is to consider I as the category of open subsets of a topological space (X, τ). By studying the fundamental notion of covering $\{U_i\}_{i \in I}$ of an open U (*i.e.*, such that $U = \bigcup_{i \in I} U_i$), Grothendieck arrives at his notion of topology, definable on an arbitrary category. Given some D, the class $J(D)$ is made up of coverings of D that will allow us to reconstruct it. In the pragmatic context we are considering, D is a neighborhood of an object and $T \in J(D)$ is a class of neighborhoods of interpretants that are enough to understand it. Below we reproduce the axioms of Grothendieck topologies and suggest their possible semiotic readings.

- $\mathcal{D}(-, D) \in J(D)$, which means that all possible interpretants are sufficient to understand the object, establishing the feasibility of the pragmatic maxim;
- if $T \in J(D)$, then $h^*(T) = \{g | \; cod(g) = E, \; hg \in T\} \in J(E)$ for any $h : E \to D$ in $\mathcal{D}$, which means that given some interpreters to understand an object, if we focus only on those that deal with some subobject, they will be enough to understand it;
- if $T \in J(D)$ and R is any sieve on D such that $h^*(R) \in J(E)$ for all $h : E \to D$ in T, then $R \in J(D)$, which we can read as if a class of interpreters R reconstructs all those that were enough to understand $F(D)$, then R is also sufficient to understand $F(D)$.

---

32 [1931–58] CP 6.104.
33 Zalamea 2012a, p. 17.
34 [1993] p. 160.
35 Zalamea 2012a, p. 16.
36 [1931–58] CP 6.104.

The generalization of these definitions to higher category theory is immediate and does not merit further discussion.[37] So now, I will be a site, that is, a (∞)-category of sign neighborhoods together with a (Grothendieck) topology that tells us how these neighborhoods are covered. Returning to Peirce's ideas, these neighborhoods must be continuous. Here a difference is marked between the approaches to the theory of sheaves given by higher category theory with respect to the traditional one. While in the latter a pre-sheaf F takes values in the category of sets, whose objects are discrete entities, in the former it does so in the (∞)-category of groupoids $\infty - \mathcal{GPD}$, which, as we saw in the previous section, are continuous structures. What's more, $\infty - \mathcal{GPD}$ is an ∞-topos, so it comes supplied with the necessary structure to make in it interesting geometric constructions (homotopy, homology, etc.). Thus we can speak of paths, equivalence of paths or convergence, which we can understand in terms of interpretation (different paths), discussion (comparing paths), and criticism (superposing paths). In this context, we can make sense of Peircean ideas of the evolution of interpretants, which affect and fuse each other, in a growing and dizzying network of divergent, convergent and parallel continuous flows.

For a map $F : I \to \infty - \mathcal{GPD}$ to be a presheaf, it must also be contravariant, that is, it must invert arrows: given $h : i \to j$, then $Fh : Fj \to Fi$. This fact of a technical nature can be read from a rather interesting aspect of Peircean semiotics. As De Tienne has observed,[38] there are two opposite directions in the semiotic process. In the first place, from the object through the sign to the successive chains of interpretants. This is the usual order of the arrows in I that we have been working with. But there is also a critical backward examination, an acknowledgment of the interpreters of the process to which they are being subjected, which adjusts and corrects the results of the interpretation to ensure its adequacy to the object. Thanks to the coupling of both views throughout the entire semiotic process, it can flow along the continuum without deviating from its purpose of representing the object.[39]

The final ingredient of the sheaf concept is the pragmatic maxim. Indeed, for all $\{U_i\}_{i \in I}$ covering of U, we can express

$$\mathcal{F}(U) = \lim\left(\prod_{i \in I} \mathcal{F}(U_i) \rightrightarrows \prod_{i,j \in I} \mathcal{F}(U_{i,j}) \substack{\to \\ \to \\ \to} \ldots \right)$$

---

[37] The interested reader can consult Lurie 2009, Section 6.2.2.

[38] De Tienne 2015.

[39] Later we will see (*Section 1.6*), that a better analogy is that of a dialogue, where each interpreter extracts information from the previous ones, to guide and promote the determination process.

The neighborhood of each object can be reconstructed by neighborhoods of interpretants as long as the latter cover the former. Thus, it is not completely necessary to work with all interpreters to fully understand a given object. Next we give an algorithm, based on the Peircean cenopythagorean categories, to gradually cover the neighborhood of the object and thus obtain a good interpretation of it:
- situate ourselves in our immediate interpretation of the object (1);
- from this, find interpretations far from ours (2), possibly in alternative, polar ways;
- via mediations (3), capture all the interpretations located between these polarities.

After executing these steps we obtain a better interpretation of the sign, however we do not reach the total interpretation. We can repeat the above process as many times as we need to improve our interpretation. Let us remember that Peirce conceived his categories as tones or tinctures on a continuum.[40] This is what we wanted to replicate in this case, which can be thought of as a "tincture game": in the first step, by locating ourselves in our initial interpretation, we are coloring the region of the continuum that this interpretation represents; from there we tincture the polarities that we can find, and finally we combine these spots in a new interpretation. Behind this fusion there is an abduction, a riddle to reconcile contradictory interpretations.

1.4 Pragmatic Maxim and Monads

In the pragmatic maxim there is a mixture of ingredients that at first glance may seem incompatible. On one hand, we have what we could call its logical part,[41] defined from a conjugation of modal operators. On the other hand, we have what we could call its behavioral part, which emphasizes the effects and circumstances. This ambivalence runs throughout Peirce's work and has been studied from different points of view. The Curry–Howard–Lambek correspondence allows us to relate both facets, starting from categorical monads:

We will use the previous correspondence to make an approximation to the pragmatic maxim that is different from what we already did in terms of limits and beams; later, in the next section, we will see that both views are equivalent. We will start by analyzing the entries in Table 1.3.

---

40 [1931–58] CP 1.353.
41 Zalamea 2012a, p. 54.

Table 1.3: Modal extension of the Curry–Howard–Lambek correspondence

Logic	Computation	Category Theory
Modal Operators	Effects/contextual computations	(co)Monads

Definition 1.4. A monad on an category $\mathcal{C}$ consists of an endofunctor $\mathbf{T} : \mathcal{C} \to \mathcal{C}$ equipped with natural transformations $\eta : id_\mathcal{C} \to \mathbf{T}$, called unit, and $\mu : \mathbf{T} \circ \mathbf{T} \to \mathbf{T}$, known as composition or multiplication, which satisfy the usual unit and associativity conditions:

$$\begin{array}{ccc} \mathbf{T}a \xrightarrow{\eta_{\mathbf{T}a}} \mathbf{TT}a & & \mathbf{TTT}a \xrightarrow{\mathbf{T}\mu_a} \mathbf{TT}a \\ \mathbf{T}\eta_a \downarrow \; \searrow^{id_{\mathbf{T}a}} \; \downarrow \mu_a & & \mu_{\mathbf{T}a} \downarrow \quad \downarrow \mu_a \\ \mathbf{TT}a \xrightarrow{\mu_a} \mathbf{T}a & & \mathbf{TT}a \xrightarrow{\mu_a} \mathbf{T}a \end{array}$$

T, which we will reserve here to refer to monads, comes from its original name, *triples*, as it is made up of three pieces of information: $\mathbf{T}, \eta, \mu$. The current nomenclature comes from the fact that they are monoids in the bicategory of categories. However, the notation **T** is extremely suggestive because, as we will see in the following pages, monads represent a Peircean Third in many categorical constructions. Also, if we want to extend Definition 1.4 to ∞-categories,[42] it will be enough to require the above diagrams to be homotopy coherent.[43] For this reason, given the limits of our approach, we will settle to work with traditional definitions.

According to the Curry–Howard–Lambek correspondence (Table 1.1, p. 2) a program in the world of computing should behave like a *pure* mathematical function, that is, have the following features: (1) running it has no observable effects, besides returning its result; (2) the outcome depends only on the value of its arguments. However, a program that behaves in this way is not very useful in the real world because if it "cannot depend on any state and cannot change any state, *i.e.*, have any input or output, then it can be argued it is the same as a program that does nothing".[44] Fortunately we can use monads to model impurities of type (1)[45] and comonads (dual definition to monads, which we will remember a little later)

---

[42] Lurie 2017, Definition 4.7.0.1
[43] That is, they can be lifted from $h\mathcal{S}$ to $\mathcal{S}$, Lurie 2009, section 1.2.6.
[44] Altenkirch and Green 2010.
[45] Related to the output changing the context, traditionally called *side-effects*, Moggi 1989

to model impurities of type (2).[46] The idea in the first situation is that when we have a program with input A and output B, instead of working with a map of the type $A \to B$, we use one of the form $A \to TB$, where the monad **T** models the effects of computation. We can use different monads to work with all kinds of effects;[47] here we are only interested in those most directly related to the Peircean "would be" (that is, an open future effect), which we already saw in a quote above (p. 5). The following fragment is more explicit:

> [...] if the same cell which was once excited, and which by some chance had happened to discharge itself along a certain path or paths, comes to get excited a second time, it is more likely to discharge itself the second time along some or all of those paths along which it had previously discharged itself than it would have been had it not so discharged itself before. This is the central principle of habit; and the striking contrast of its modality to that of any mechanical law is most significant. The laws of physics know nothing of tendencies or probabilities; whatever they require at all they require absolutely and without fail, and they are never disobeyed. Were the tendency to take habits replaced by an absolute requirement that the cell should discharge itself always in the same way, or according to any rigidly fixed condition whatever, all possibility of habit developing into intelligence would be cut off at the outset; the virtue of Thirdness would be absent.[48]

In the following examples we will see how some monads fit the ideas expressed in the previous quote.

Example 1.1.
1. One of the simplest is the *maybe*, which on the category of sets is defined as
   - $TA = A + \{*\}$, the coproduct of A with a singleton;
   - $\eta_A = A \xrightarrow{inc_A} A + \{*\}$;
   - $\mu_A = (A + \{*\}) + \{*\} \xrightarrow{proj_{A+\{*\}}} A + \{*\}$.

   As a program $P : A \to B$ in a functional language must always return an output in the range for any input in the domain, we can introduce the possibility of computational failure by working with $P : A \to TB$, where the point of $\{*\}$ is interpreted as "no value returned".
2. The *would be* is related to a notion of "chance". Lawvere observed[49] that probability theory can be introduced into category theory (and hence in computation) by means of monads. The idea is to assign to each "spaces of out-

---

46 *Contextual computations*, that depend upon, or make demands on, the context of evaluation. See Uustalu and Vene 2008.
47 See for example, Moggi 1989
48 **[1931–58]** CP 1.390.
49 Lawvere 1962

comes" X a space TX of "laws of random outcomes on X". Depending on the base category (sets, measurable spaces, topological spaces, locales, etc.), different types of monads can be defined that make use of the underlying structure to make richer probability theories. Perhaps the simplest case is when X is simply a set, and the monad has the form:

- TX, the set whose elements are functions $p : X \to [0, 1]$ such that $p(x) \neq 0$ for only finite x, and $\sum_{x \in X} p(x) = 1$.
- $\eta_X : X \to TX$, the map which sends element $x \in X$ to the function $\eta_x : X \to [0, 1]$, given by

$$\eta_x(y) = \begin{cases} 1, & \text{if } y = x; \\ 0, & \text{if } y \neq x. \end{cases}$$

- $\mu_X : TTX \to TX$, the map which sends $e \in TTX = TX \to [0, 1]$ to $\mu_X(e) : X \to [0, 1]$, given for $y \in X$ by

$$\mu_X(e)(y) = \sum_{p \in TX} p(y) e(p).$$

3. Let (M, e, m) be a monoid with identity element e and operation $m : M \times M \to M$. The output monad (also known as writer monad) is given by
   - $TA := A \times M$
   - $\eta_A := A \xrightarrow{inc_A} A \times \{*\} \xrightarrow{id_A \times e} A \times M$
   - $\mu_A := (A \times M) \times M \xrightarrow{asoc} A \times (M \times M) \xrightarrow{id_A \times m} A \times M$.

We can think of this monad as keeping track of the times the program has been run (or the times the cell has been excited, in the Peircean analogy).

Definition 1.5. Let (T, η, μ) be a monad on a category $\mathcal{C}$. A T-algebra on this monad is a pair (a, h) where $a \in Ob(\mathcal{C})$, $h : Ta \to a \in Mor(\mathcal{C})$ such that

$$\begin{array}{ccc} a \xrightarrow{\eta_a} Ta & & TTa \xrightarrow{Th} Ta \\ {}_{id_a} \searrow \downarrow h & & {}_{\mu_a} \downarrow \quad \downarrow h \\ a & & Ta \xrightarrow{h} a \end{array}$$

Proposition 1.3. Given a monad (T, η, μ), we can form a new category $\mathcal{C}^T$, called the Eilenberg–Moore category, consisting of:
- objects: the T-algebras,

- morphisms: given two algebras (a, h), (b, k), a morphism $f : (a, h) \to (b, k)$ is a $f : a \to b \in Mor(\mathcal{C})$ such that

$$\begin{array}{ccc} Ta & \xrightarrow{Tf} & Tb \\ h \downarrow & & \downarrow k \\ a & \xrightarrow{f} & b \end{array}$$

is commutative.

Proposition 1.4. The forgetful functor

$$\mathcal{C}^T \xrightarrow{U^T} \mathcal{C}$$

$$\begin{array}{ccc} Ta \xrightarrow{h} a & & a \\ Tf \downarrow \quad \downarrow f & \mapsto & \downarrow f \\ Tb \xrightarrow{k} b & & b \end{array}$$

has a left adjoint

$$\mathcal{C} \xrightarrow{F^T} \mathcal{C}^T$$

$$\begin{array}{ccc} a & & Ta \xleftarrow{\mu_a} TTa \\ h \downarrow & \mapsto & Th \downarrow \qquad \downarrow TTh \\ b & & Tb \xleftarrow{\mu_b} TTb \end{array}$$

and besides $U^T \circ F^T \simeq \mathbf{T}$.

Remark 1.1. We have already mentioned (p. 16) how colimits correspond to the ultimate intellectual interpreter (habit) in infinite semiosis. On the other hand, the quotation [1931–58] CP 1.390 shows how the effects generate the habit and how later the habit will generate corresponding effects. Thinking $\mathcal{C}^T$ as the category of effects associated with $\mathcal{C}$, the good correspondence of pragmatism with category theory would require that F^T commute with colimits; which is guaranteed by the existence of its right adjoint U^T.

Proposition 1.5. Let $L : \mathcal{C} \leftrightarrows \mathcal{D} : R$ be a pair of adjoint functors, with unit $\eta : id_\mathcal{C} \Rightarrow R \circ L$ and counit $\epsilon : L \circ R \Rightarrow id_\mathcal{D}$. Then the triple $(T = R \circ L, \eta, R(\epsilon_{L(\_)}))$ yields a monad on $\mathcal{C}$.

The previous results show the strong relationship between monads and adjoints. In particular, Proposition 1.5 tells us that every pair of adjoint functors generates a monad. If we think of an adjunction as a kind of "dialectical pendulum", an application of **T** would be an oscillation of that pendulum. The pendulum is one of the most beautiful and accurate images to capture Peircean thought in general.[50] The trajectory that goes from x to Tx crosses the frontier of $\mathcal{C}$ back and forth on its journey using the pair of adjoint functors, and thanks to this going and coming across the border we increase our understanding of the object.[51] In this way, we can also think of $\mathcal{C}^T$ as a category of pendular weavings, where the axioms of **T**-algebras (Definition 1.5) provide operations that allow us to navigate between the different trajectories: η retracts oscillations, while μ composes them.

Definition 1.6. Let $F : \mathcal{B} \leftrightarrows \mathcal{C} : U$ be a pair of adjoint functors, with counit $\epsilon : FU \Rightarrow id_{\mathcal{C}}$, and **T** the associated monad. The canonical **T**-algebra functor K^T is defined by

$$
\begin{array}{ccc}
\mathcal{C} & \xrightarrow{K^T} & \mathcal{B}^T \\
a & & Ua \xleftarrow{U\epsilon_a} UFUa \\
g\downarrow & \mapsto & Ug\downarrow \qquad \downarrow UFUg \\
b & & Ub \xleftarrow[U\epsilon_b]{} UFUb
\end{array}
$$

The functor U will be called monadic (resp. premonadic) if K^T is an equivalence of categories (resp. full and faithful).

As will be clear from the discussion in the next section, it is convenient to denote the category $\mathcal{B}^T$ by $\mathcal{MDES}(U)$ and call it the category of monadic descent data of U. If U^T, F^T are the functors of Proposition 1.4, by construction we have that in the diagram

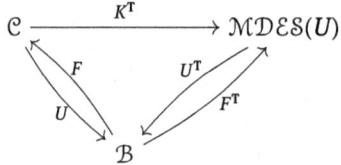

$U^T K^T \simeq U$ and $K^T F \simeq F^T$.

All the previous constructions can be dualized directly, however we think it is convenient to make some of them explicit to fix the notation that we will use later.

---

[50] For an in-depth philosophical study of this analogy, see Zalamea 2010c, especially chapter 1.
[51] Zalamea 2010c p. 32.

Definition 1.7. A comonad on an category $\mathcal{C}$ consists of an endofunctor $\mathbf{D} : \mathcal{C} \to \mathcal{C}$ equipped with maps $\varepsilon : \mathbf{D} \to id_{\mathcal{C}}$ and $\delta : \mathbf{D} \to \mathbf{D} \circ \mathbf{D}$ which satisfy

$$\begin{array}{ccc} DX \xrightarrow{\delta_X} DDX & \quad & DX \xrightarrow{\delta_X} DDX \\ \delta_X \downarrow \searrow^{id_{DX}} \downarrow \varepsilon_{DX} & & \delta_X \downarrow \qquad \downarrow D\delta_X \\ DDX \xrightarrow[D\varepsilon_X]{} DX & & DDX \xrightarrow[\delta_{DX}]{} DDDX \end{array}$$

Currently, programs are executed on a large number of devices (PCs, servers, phones, smart glasses, etc.). Context-aware programming seeks to design programs that are aware of the environment where they are running, to take advantage of the various unique capabilities of each of them (access to GPS or augmented reality, different capabilities of graphics or RAM cards, etc.). In pragmatic terms, programs are signs that behave in different ways depending on the context where they are interpreted/executed. From the work of Uustalu and Vene[52] contextual computation is modeled by comonads. Here we will see just a couple of extremely simple examples; for many more interesting ones, we recommend the reader to consult the works of Orchard and Petricek.[53]

Example 1.2.
1. Given a fixed object C of $\mathcal{C}$, the product comonad is defined by
   - $DA := A \times C$
   - $\varepsilon_A := A \times C \xrightarrow{\pi_A} A$, given by the projection of the product to A
   - $\delta_A := A \times C \xrightarrow{\langle id_{A \times C}, \pi_C \rangle} (A \times C) \times C$, obtained by the universal property of the product.

   The idea is to see C as a type of contexts. A program $f : A \to B$ only takes into account the inputs in A, while one $\check{f} : A \times C \to B$ also considers the different contexts C in which the program runs.

2. Let (M, e, m) be a monoid with identity element e and operation $m : M \times M \to M$. The exponent comonad is given by
   - $DA := M \Rightarrow A$
   - $\varepsilon_A := (M \Rightarrow A) \xrightarrow{inc} (M \Rightarrow A) \times \{*\} \xrightarrow{id \times e} (M \Rightarrow A) \times M \xrightarrow{ev} A$
   - $\delta_A(f, s, t) := f(m(s, t))$, where $f : M \Rightarrow A$, $s, t \in M$.

   Now the context is a monoid, say the natural numbers $\mathbb{N}$. We have already seen in Example 1.1 that we can keep a count of the times a program has

---
[52] Uustalu and Vene 2008.
[53] Orchard 2014, Petricek 2017.

been run (or a cell excited). This comonad takes this number into account to produce the output: the larger the number, the greater the appearance of a desired value. This is the spirit of the quote [1931–58] CP 1.390 that we reproduced on p. 23.

The next class of monads will play a key role in what follows.

Definition 1.8. An *idempotent monad* is a monad (T, μ, η) on a category $\mathcal{C}$ such that $\mu : TT \to T$ is a natural isomorphism.

An idempotent monad behaves very similarly to the possibility operator $\Diamond$, while an idempotent comonad behaves like the necessity operator $\Box$. More precisely we have:

Proposition 1.6. Let $\Box$ be a comonad idempotent and $\Diamond$ a monad idempotent such that $\Diamond \dashv \Box$. Then the following properties are valid

1. $K : \Box(p \to q) \to (\Box p \to \Box q)$
2. $T : \Box p \to p$
3. $4 : \Box p \to \Box\Box p$
4. $B : p \to \Box\Diamond p$
5. $D : \Box p \to \Diamond p$
6. $4.2 : \Diamond\Box p \to \Box\Diamond p$
7. $5 : \Diamond p \to \Box\Diamond p$

Proof.
1. Endofunctor $\Box$.
2. Counit of comonad $\Box$.
3. Idempotency of $\Box$.
4. Unit of $\Diamond \dashv \Box$.
5. Composition of the counit of the comonad ($\varepsilon : \Box p \to p$) with the unit of the monad ($\eta : p \to \Diamond p$).
6. Composition of the counit ($\Diamond\Box p \to p$) and unit ($p \to \Diamond\Diamond p$) of $\Diamond \dashv \Box$.
7. Application of the definition of adjunction to the idempotency of $\Diamond$ ($\Diamond\Diamond p \to \Diamond p$) $\leftrightarrow$ ($\Diamond p \to \Box\Diamond p$).

□

The inverse is partially true: a propositional (first order) modal logic where axioms analogous to the previous ones hold gives rise to an adjunction of idempotent (co)monads on the Lindenbaum algebra (syntactic category).[54] Also, all (co)monads can be "completed" to behave like modal operators:

---

[54] See for example Reyes and Zawadowski 1993.

Proposition 1.7. Let $\mathcal{C}$ be a complete, well-powered category, $\mathcal{MON}(\mathcal{C})$ be the category of monads on $\mathcal{C}$ and $\mathcal{IMON}(\mathcal{C})$ be the category of idempotent monads on $\mathcal{C}$. Then the inclusion $i : \mathcal{IMON}(\mathcal{C}) \hookrightarrow \mathcal{MON}(\mathcal{C})$ has a right adjoint $q : \mathcal{MON}(\mathcal{C}) \to \mathcal{IMON}(\mathcal{C})$.[55]

In this way, we finish our purpose of reviewing the equivalence of entries in Table 1.3: (co)monads correspond both to modal operators in the world of logic and to effects/contexts in the world of computing. As we mentioned in *Section 1.1*,[56] we have a notion of "process" or "transit" that is embodied in the world of logic in "proofs", in the world of computation in "programs", in the theory of categories in "functors", and in pragmatism in "signs". Now, linked to the "transit", there must be a notion of "resistance", which in a certain sense opposes the plasticity of the transit and allows us to grasp it. In pragmatism the "effects" are the signals left by the transit of semiosis, like "paths" in the interpreter, which over time become habits. Following these traces is how we can understand the object by means of a sort of maximal pragmatics. In computing, the "side effects" are what make a program useful. In existential graphs, modal operators arise from resistance to iteration through a broken cut, and the different permeability of the cut is what allows the different modal systems to be distinguished (see *Subsection 3.1.3* in this volume). In category theory, as we will see in *Section 1.6*, monads are what allow us to grasp physical reality.

1.5 Pragmatic Maxim and Descent Theory

In this section we will see how the approach to the maxim in terms of sheaves (which allow us to reintegrate the local visions of the different interpreters into a global understanding of the object) fits with monads (which introduce the modal character of the maxim and allow us to re-understand it in terms of effects/circumstances). The ideas arise from descent theory, developed by Grothendieck between the late 1950's and early 1960's.[57] The fundamental observation that descent theory can be rewritten in terms of monads is due to Bénabou and Roubaud.[58]

---

[55] Fakir 1970.
[56] See p. 6 above.
[57] Grothendieck 1959, Grothendieck and Raynaud 1971, Chapters VI–IX.
[58] Bénabou and Roubaud 1970. For an alternative display in a less technical context we recommend Janelidze and Tholen 1994. Anyway, we will do a quick discussion to contextualize the reader unfamiliar with this theory.

The motivation of sheaf and descent theory is to glue local structures to obtain global ones. However, there are some differences in approach. In sheaf theory we are more interested in direct comparisons (except equality) while in descent theory we work with indirect comparisons (except equivalence). To guide our further discussion it is useful to clarify this point. Let $\mathcal{O}$ be the set of opens of a topological space (X, τ) ordered by inclusion and consider another topological space C. We can define the functor

$$\mathcal{O} \xrightarrow{F} \mathcal{SET}$$
$$U \longmapsto \{f : U \to C, \text{ continuous}\}.$$

Saying that F is a sheaf means that given a $\{U_i\}_{i \in I}$ covering of an open U (i.e., such that $U = \bigcup_{i \in I} U_i$) and a family of local functions $\{f_i : U_i \to C \in F(U_i)\}$ subject to a good *compatibility condition*

$$\forall i, j \in I, \ f_i|_{U_i \cap U_j} = f_j|_{U_i \cap U_j}, \tag{1.1}$$

the local functions can be glued into a single global function $f : U \to C \in F(U)$. The point is that in the compatibility condition we compare via *equality*. Let us now consider a slightly different problem. $\{U_i\}_{i \in I}$ continues to be a covering of U but now the family of functions that we want to glue are of the form $\{f_i : C_i \to U_i\}$ to obtain the global $f : C \to U$. Denoting with $\mathcal{TOP}/X$ the comma category over a topological space X, the key point is that the above functions live in different categories: $f_i \in \mathcal{TOP}/U_i$ and $f \in \mathcal{TOP}/U$. Thus, to compare them we need functors. At this point it is convenient to remember that every function $h : V \to W \in \mathcal{TOP}$ gives rise to two natural functors between the induced comma categories, $h_!$ (the composite with h) left adjoint to h^* (the pullback along h):

$$\begin{array}{ccc} \mathcal{O} & \xrightarrow{\mathcal{F}} & \mathcal{CAT} \\ V & & \mathcal{TOP}/V \\ h \downarrow & \longmapsto & h_! \,\bigcap\, h^* \\ W & & \mathcal{TOP}/W \end{array} \tag{1.2}$$

Since pullbacks are limits, when we compare information obtained by them this will be accurate only except isomorphisms.

Remark 1.2. In the passage from the rigid definition of identity through equality, to more flexible ones, first by isomorphism and then by continuous deformations, lies one of the main contributions of (higher) category theory to philosophy. Many thinkers[59] have noted how identity conceived as equality is too static and that the identity of an evolving being requires other approaches. In this sense, Peirce wrote:

> To say that who commanded the French in the battle of Leipzig commanded them in the final battle of Waterloo, is not merely a statement of identity: it is a statement of Becoming. There is an existential continuity in time between the two events. But so understood the statement asserts no Significative Identity, inasmuch as the intervening continuum is a continuum of Assertoric Truths. Now, upon a continuous line there are no points (where the line is continuous). There is only room for points—possibilities of points. Yet it is through that continuum, that line of generalization of possibilities, that the actual point at one extremity necessarily leads to the actual point at the other extremity.[60]

This conception of identity as a continuum evolving on another continuum reaches its technical realization in the line of identity in existential graphs. More generally, the different disciplines in the Curry–Howard correspondence (Tables 1.1, 1.2), have arrived at analogous *ideas* for apparently different motivations, but originated by the same dialectic of *equality* and *difference*: how to recognize the identity of a being through its successive transformations? Thus, for example, the need to establish an identity definition with good computational properties, led researchers in programming languages to understand a type as a ∞-groupoid, in such a way that the identity type between its terms can be seen as a class of continuous paths that vary over it.

The following definition abstracts the situation we are discussing.

Definition 1.9. A pseudo-functor $\mathcal{F}$ on $\mathcal{C}$ consists of
1. for each object U of $\mathcal{C}$ a category $\mathcal{F}U$;
2. for each arrow $f : U \to V$ a functor $f^* : \mathcal{F}V \to \mathcal{F}U$;
3. for each object U of $\mathcal{C}$ an isomorphism $\beta_U : id_U^* \simeq id_{\mathcal{F}U}$ of functors $\mathcal{F}U \to \mathcal{F}U$;
4. for each pair of arrows $U \xrightarrow{f} V \xrightarrow{g} W$ an isomorphism

$$\alpha_{f,g} : f^*g^* \simeq (gf)^* : \mathcal{F}W \to \mathcal{F}U.$$

These data must satisfy the appropriate conditions of identity and associativity except for natural isomorphisms. Furthermore, for each $f : U \to V$ we require a functor $f_! : \mathcal{F}U \to \mathcal{F}V$, such that $f_! \dashv f^*$.

---

59 For example, Florensky 1997, pp. 59–61.
60 [1976] NEM 4.330.

Remark 1.3. We take the opportunity to introduce a small notation. All commutative diagrams

$$\begin{array}{ccc} A & \xrightarrow{a} & B \\ c\downarrow & & \downarrow b \\ C & \xrightarrow{d} & D \end{array}$$

give rise to natural isomorpisms

$$\tau_{a,b,c,d} : c^*d^* \xrightarrow{\alpha_{c,d}} (dc = ba)^* \xrightarrow{\alpha^{-1}_{a,b}} a^*b^*$$

between functors from $\mathcal{F}(D)$ to $\mathcal{F}(A)$. We will simply write τ when the occurrence of $\{a, b, c, d\}$ is clear from the context.

As we have already mentioned, the compatibility condition must be expressed by means of pullbacks. For this we write the diagrams

$$\begin{array}{ccc} g_i^*(U_i \cap U_j) & \xrightarrow{p_1^i} & U_i \cap U_j \\ \downarrow & & \downarrow \\ C_i & \xrightarrow{g_i} & U_i \end{array} \qquad \begin{array}{ccc} g_j^*(U_i \cap U_j) & \xrightarrow{p_2^j} & U_i \cap U_j \\ \downarrow & & \downarrow \\ C_j & \xrightarrow{g_j} & U_j \end{array} \quad (1.3)$$

and we compare the information with the morphisms given by the universal property

$$\begin{array}{ccc} g_i^*(U_i \cap U_j) & \xrightarrow{m_{i,j}} & g_j^*(U_i \cap U_j) \\ & \searrow_{p_1'} \quad \swarrow_{p_2'} & \\ & U_i \cap U_j & \end{array}$$

We want to rewrite this information without mentioning the $\{U_i\}$ cover. For this we define $E = \coprod_{i \in I} U_i$, and the functions $p : E \to U$, given by $p(i, u) = u$, and $g : C \to E$, given by the disjoint union of the $\{g_i : C_i \to U_i\}_{i \in I}$. Thus, if we take the pullback

$$\begin{array}{ccc} E \times_U E & \xrightarrow{p_1} & E \\ p_2 \downarrow & & \downarrow p \\ E & \xrightarrow{p} & U \end{array} \quad (1.4)$$

we see, thanks to the fact that it is a disjoint union, that $E \times_U E = \coprod_{i,j \in I} U_i \cap U_j$. Therefore, we can also make the union on i, j of the $m_{i,j} : g_i^*(U_i \cap U_j) \to g_j^*(U_i \cap U_j)$ to get the morphism

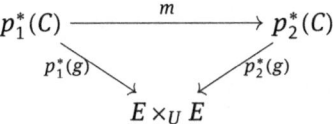

Let's observe that thanks to the morphisms $g : C \to E$, $p_1^*(g) : p_1^*(C) \to E \times_U E$, $p_2^*(g) : p_2^*(C) \to E \times_U E$ we have $C \in Ob(\mathcal{TOP}/E)$ and $p_1^*(C), p_2^*(C) \in Ob(\mathcal{TOP}/E \times_U E)$. To say that m is made up of isomorphisms, we consider the pullback

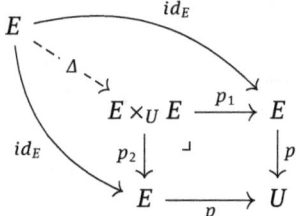

and demand that $IC_2^{-1} \circ \Delta^*(H) \circ IC_1 = id_C$:[61]

$$
\begin{array}{ccc}
\mathcal{F}(E \times_U E) & \xrightarrow{\Delta^*} & \mathcal{F}E \\
p_1^*C & & \Delta^*p_1^*C \quad \overset{IC_1}{\leftarrow} \\
m \downarrow \quad \mapsto & & \Delta^*m \downarrow \qquad\qquad C \\
p_2^*C & & \Delta^*p_2^*C \quad \overset{IC_2}{\leftarrow}
\end{array}
$$
(1.5)

where $IC_i = \alpha_{\Delta,p_i}(C)^{-1}\beta_U(C)^{-1}$ (α, β as in Definition 1.9).

Let's now look at another difference between sheaf and descent theory. The compatibility condition given by equality (1.1) is sufficient to guarantee the local gluing in the case of sheaves. When compatibility is thought through equivalence, the price to pay for a weaker condition is the need for an additional one that tells us that the family of isomorphisms $m_{i,j}$ is coherent. For this we denote $U_{ijk} :=$

---

[61] Grothendieck 1959, p. 302.

$U_i \cap U_j \cap U_k$ and m_{ij}^k the restriction of m_{ij} to $g_i^*(U_{ijk})$. The *cocycle condition* requires for all $i, j, k \in I$ the following diagram to be commutative:

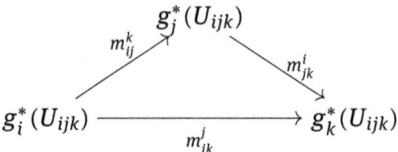

To generalize this condition to bi-pseudofunctors, let's consider pullbacks:

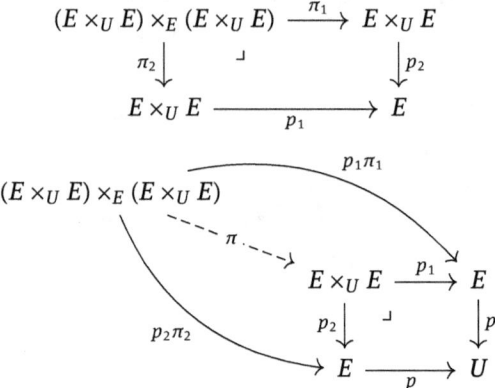

and we demand that the hexagon

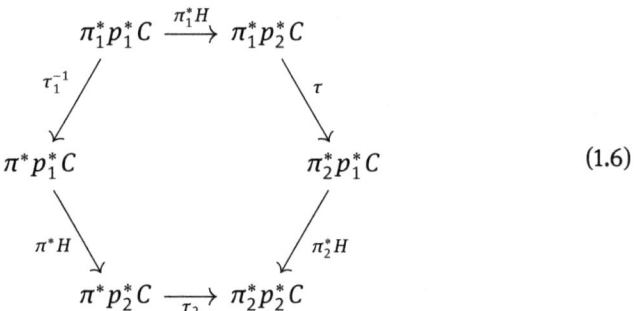 (1.6)

is commutative. Here the isomorphisms τ are given as in the remark above. Since m is also an isomorphism, so are all arrows in the hexagon.

Definition 1.10. Given a pseudofuntor $\mathcal{F}$ over a category with pullbacks $\mathcal{C}$, the category of geometric descent data relative to a morphism $p : E \to U$, denoted $\mathcal{GDES}(p)$, consists of:

1. objects: pairs (c, m) such that $c \in \mathcal{F}(E)$, $m : p_1^*c \to p_2^*c$, that validate the conditions indicated in diagrams (1.5), (1.6).
2. morphisms: $f : (c, m) \to (c', m')$ such that $f : c \to c' \in Mor(\mathcal{F}(E))$ and

$$\begin{array}{ccc} p_1^*c & \xrightarrow{m} & p_2^*c \\ p_1^*f \downarrow & & \downarrow p_2^*f \\ p_1^*c' & \xrightarrow{m'} & p_2^*c' \end{array}$$

commutes.

Remark 1.4. The natural isomorphism τ applied to diagram (1.3) induces a functor

$$\begin{array}{ccc} \mathcal{F}(U) & \xrightarrow{\tau} & \mathcal{GDES}(p^*) \end{array}$$

$$\begin{array}{ccc} b & & p_1^*p^*b \xrightarrow{\tau(b)} p_2^*p^*b \\ f \downarrow & \mapsto & p_1^*p^*f \downarrow \qquad\qquad \downarrow p_2^*p^*f \\ b' & & p_1^*p^*b' \xrightarrow{\tau(b')} p_2^*p^*b' \end{array}$$

The pair $(p^*b, \tau(b))$ satisfies the conditions of Definition 1.10 thanks to the naturality of τ. Thus, by construction, the diagram

$$\begin{array}{ccc} \mathcal{F}(U) & \xrightarrow{\tau} & \mathcal{GDES}(p^*) \\ & p^* \searrow \quad \swarrow M & \\ & \mathcal{F}(E) & \end{array}$$

is commutative, where M is the obvious forgetful functor.

We now come to a last difference between sheaf and descent theory: the first requires from the beginning that the structures in play *actually* satisfy a glueing property, while the second deals with conditions on a morphism p to make *possible* a glueing.

Definition 1.11. A morphism p is called (effective) $\mathcal{F}$ descent morphism if the above τ is full and faithful (an equivalence of categories).

Comparing this diagram with that of Definition 1.6 it is natural to ask if $\mathcal{GDES}(p^*)$ is equivalent to $\mathcal{MDES}(p^*)$. For this to happen, all we need is a good behavior of the adjunction $(\_)_! \dashv (\_)^*$.

Definition 1.12. A bipseudofunctor $\mathcal{F}$ satisfies the *Beck–Chevalley property* if for all pullbacks

$$\begin{array}{ccc} A & \xrightarrow{s} & B \\ t\downarrow & \lrcorner & \downarrow u \\ C & \xrightarrow{v} & D \end{array}$$

its associated diagram by $\mathcal{F}$

$$\begin{array}{ccc} \mathcal{F}A & \xrightarrow{s_!} & \mathcal{F}B \\ t^*\uparrow & & \uparrow u^* \\ \mathcal{F}C & \xrightarrow{v_!} & \mathcal{F}D \end{array}$$

induces a natural isomorphism $\theta : s_! \circ t^* \xrightarrow{\sim} u^* \circ v_!$.

Proposition 1.8 (Bénabou–Roubaud Theorem)**.** Assuming the Beck–Chevalley condition, there is an equivalence of categories $\delta : \mathcal{GDES}(p^*) \xrightarrow{\sim} \mathcal{MDES}(p^*)$ such that the following diagram is commutative:

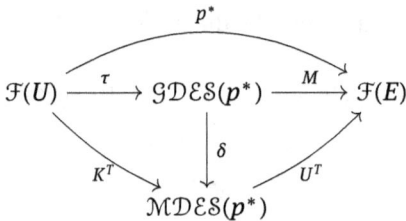

Proof. We will only mention how to transfer the data. Thanks to the adjunction $p_{2!} \dashv p_2^*$ we have natural isomorphisms

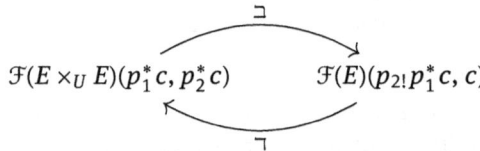

Applying Beck-Chevalley to the pullback (1.3) we obtain $\theta_c : p_{2!}p_1^*c \to p^*p_!c$, which induces a new pair of natural isomorphisms:

$$\begin{array}{ll} \mathcal{F}(E \times_U E)(p_1^*c, p_2^*c) & \mathcal{F}(E)(p_!p^*c, c) \\ m \longmapsto & (\sqsupset m) \circ \theta_c^{-1} \\ \sqsupset(h \circ \theta_c) \longleftarrow & \longmapsto h \end{array}$$

$\square$

Thanks to the previous equivalence, diagrams (1.5), (1.6) correspond to the first and second diagrams of Definition 1.5, respectively. As it is natural to expect for a reconstruction to be valid for a morphism $p : E \to U$, it must "cover" U in a suitable way. What we mean by this depends on the category in which we are working. In the case of comma categories on $\mathcal{TOP}$ (the initial example on p. 30), among the classes of maps for which effective descent is valid are open surjections, proper maps, locally sectionable maps (in particular surjective local homeomorphisms).[62]

Finally, we will read all these categorical results in terms of the pragmatic maxim. U plays the role of the *object of our conception* or *intellectual concept* of the statements of the maxim (*Section 1.3*), $\{U_i\}_i$ is a family of its interpretants and E is only the disjoint union of these. As points, they are ideal entities and we can only apprehend the neighborhoods that they give rise to in a continuum via the functor $\mathcal{F}$. In this continuum we can speak of $\mathcal{F}U$ as a global entity and of $\mathcal{F}E$ as a family of local entities. $\mathcal{GDES}(p^*)$ is a category made up of the local invariants obtained by contrasting the interpretants. If the neighborhoods of the interpretants adequately cover the neighborhood of the object, then these invariants are sufficient to reconstruct the object. According to Proposition 1.8 this procedure is equivalent to considering $\mathcal{MDES}(p^*)$, which is, by definition, the category $\mathcal{F}(E)^\mathbf{T}$ of effects of the family of interpreters. In this case, for the reconstruction to be possible, we need, according to Definition 1.6, the functor p^* to be monadic. In category theory, to guarantee this property, one of the different versions of Beck's theorem is used, named in honor of the mathematician who discovered it in the mid-sixties.[63] In its most general form, it states:

Proposition 1.9. Let $F : \mathcal{A} \to \mathcal{B}$ be a functor. The following conditions are equivalent:

---

[62] Janelidze and Tholen 1994, p. 274.
[63] Beck 1967.

- F is monadic.
- 1. F has a left ajoint;
- 2. F reflects isomorphisms (*i.e.*, if $F(f)$ is an isomorphism then f is an isomorphism);
- 3. If a pair $u, v : X \rightrightarrows Y$ in $\mathcal{A}$ is such that $(F(u), F(v))$ has a split coequalizer in $\mathcal{B}$, then (u, v) has a coequalizer in $\mathcal{A}$ which is preserved by F.

In our case, we have already seen that $p_!$ is the left adjoint of p^* (for a pragmatic discussion of this fact, see Remark p. 25). For the other two conditions, let us remember that the co-equalizers measure the degree of distinguishability of two parallel arrows $A \rightrightarrows B$: it is a quotient object $q : B \twoheadrightarrow Q$ which is large if the arrows are very similar and small in the opposite case. In precise terms (2) and (3) express what we could call *pragmatic distinctness*, *i.e.*, that "every real distinction of thought, no matter how subtle it may be" must be "tangible and conceivably practical",[64] that is, established thanks to the different effects it gives rise to.

Let's finally observe that the reconstruction of the maxim that we offered in this section was possible thanks to the pendulum (back-and-forth) information flow between the different levels of $\mathcal{F}$ given by the adjointness $(\_)_! \dashv (\_)^*$. In the next section we will study this process in more detail.

1.6 Determination Process and Adjoint Strings

In this section we would like to study how information evolves in peircean semiotics. As we mentioned earlier, this process does not go univocally from the object to the interpreter, but rather is pendular, with constant comings and goings. The following quote, where Peirce relates one of his experiments, is one of the best explanations of his thought.[65]

> [...] I am to oscillate simultaneously a yard reversible pendulum and a metre reversible pendulum. I shall thus ascertain with great precision the ratio of their lengths [...] The knife-edges of the pendulums will be interchanged and the experiments repeated. Finally, the yard pendulum will be compared with a yard bar and the metre pendulum with a metre bar, and last of all the yard pendulum with its yard bar will be sent to England, the metre pendulum with its metre bar to France, for comparison with the primary standards; and thus it is believed the ratio of yard to metre will be ascertained with the highest present attainable exactitude.[66]

---

64 [1931–58] CP 5.400.
65 Zalamea 2010c p. 23.
66 [1931–58] CP 7.16, "Report of a Conference on Gravity Determinations".

Let us denote with X the entity that we want to determine, in this case the ratio between the meter and the yard. Since X lives in a true continuum, by inextensibility[67] we cannot determine its exact value, but only neighborhoods that contain it. The simplest way to find these neighborhoods is to establish, via a double pendulum process, an interval that delimits the value of X above and below. For this reason, we need at least two adjunctions in the determination process.

Definition 1.13. Given a triad of functors

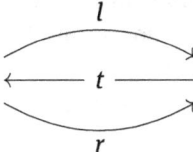

we will say that they form an adjoint triple if $l \dashv t$ and $t \dashv r$, which we denote more compactly as $l \dashv t \dashv r$.

An adjoint triple induces two adjunctions of (co)monads. First, we have $tl \dashv tr$, where tl is a monad and tr is a comonad. Second, we have $lt \dashv rt$, where rt is a monad and lt is a comonad. By requiring simple properties on one of the functors of the triple we can make one of these adjunctions become one of idempotent (co)monads (that is, of modal operators), while the other is trivialized. Throughout this section we will constantly use the notation of Definitions 1.4, 1.7, where we reserve **T** and **D** respectively for monads and comonads.

Definition 1.14. An adjoint triple $l \dashv t \dashv r$ is of the form
1. **T** $\dashv$ **D** if t is full and faithful.

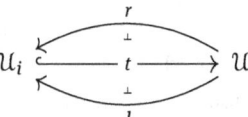

2. **D** $\dashv$ **T** if any of the outer functors r or l is full and faithful.

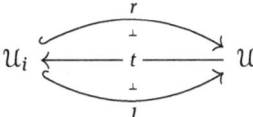

---

67 Zalamea 2012a, p. 17.

A classic result of category theory states that given an adjoint $F \dashv U$, saying that U (resp. F) is full and faithful is the same as saying that $\varepsilon : FU \to id$ (resp. $\eta : id \to UF$) is an isomorphism. Thus in the case $\mathbf{T} \dashv \mathbf{D}$ the adjunction $lt \dashv rt$ is trivialized (that is, $lt \simeq rt \simeq id$), while $tl \dashv tr$ becomes one of an idempotent monad left adjoint to an idempotent comonad ($tltl \simeq tl$ and $trtr \simeq tr$). In this case t is an embedding, so $\mathcal{U}_i$ is a subcategory of $\mathcal{U}$. Another result of folklore states that r is full and faithful iff l is too. Thus, in the case $\mathbf{D} \dashv \mathbf{T}$, $tr \simeq id$ and $tl \simeq id$, that is, the adjunction $tl \dashv tr$ is trivialized. Also, $lt \dashv rt$ becomes one of an idempotent comonad left adjoint to an idempotent monad. Then $tr \simeq id$ and $tl \simeq id$ implies that both r and l are embeddings, so we can think that in $\mathcal{U}$ there are two copies of $\mathcal{U}_i$ unified by the projection t, forming a kind of cylinder.

To better understand these two forms of determination, let's refer to Peirce:

> [...] signs require at least two Quasi-minds; a *Quasi-utterer* and a *Quasi-interpreter*; and although these two are at one (i.e., are one mind) in the sign itself, they must nevertheless be distinct. In the Sign they are, so to say, welded. Accordingly, it is not merely a fact of human Psychology, but a necessity of Logic, that every logical evolution of thought should be dialogic.[68]

What is most striking about the passage is the distinction of the two Quasi-minds despite being welded in the sign. The function of this separation is mainly to distinguish the two ways in which the determination process can be carried out. On one hand, we have the objectively general signs, where it is left *to the interpreter the right of completing the determination for himself*.[69] On the other, we have the objectively vague ones, where *the right of determination is not distinctly extended to the interpreter [but] it remains the right of the utterer*.[70] To see how this connects with adjunction of (co)monads theory, let's denote with $\mathcal{U}$, $\mathcal{U}_i$ the *Quasi-utterer* and the *Quasi-interpreter* categories, respectively. Let us consider an object c of $\mathcal{U}$, which will play the role of sign to be determined. We must emphasize that we should think of this sign as an ideal entity (a point on a continuum), which we can only conceive by neighborhoods. In the case of an adjoint of the form $\mathbf{T} \dashv \mathbf{D}$ the application of the various functors on c gives rise to the diagram

---

68 [1931–58] CP 4.551.
69 [1931–58] CP 5.505, 5.448.
70 [1931–58] CP 5.506.

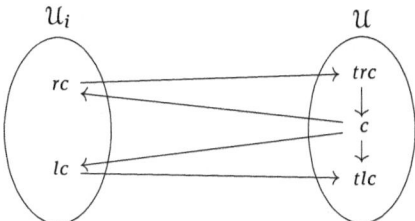

In this case the determination interval is originated in $\mathcal{U}_i$, so we could call this a general process. Note that by the rules $lt \simeq rt \simeq id$, even if we re-apply some of the functors, there will be no changes in the diagram, that is, there are no more possible determinations with this same triple of adjunctions. In the case of an adjoint of the form **D ⊣ T** the diagram is

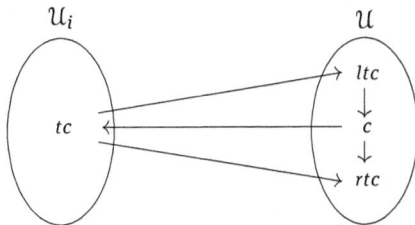

In this case the determination interval is defined in $\mathcal{U}$, so we could call this process vague. Here since $tr \simeq tl \simeq id$ there is no more determination with the same triple of functors.

In both cases, the neighborhood of delimitation takes the form of an interval with the filtrations of the entity by the modalities as extremes. We can think of these intervals as a transit from the first modality, the entity, to the second, in the direction of the arrows, passing through all the objects that factor them. In Peircean thought the transits of possibility, actuality and necessity modalities play an important role, for example in determining the continuum. As we will see in Examples 1.3 and 1.4, this same role is played by the transit of categorical modalities.

Example 1.3. To see concretely the connection of indeterminacy and adjoint triples, let us review Lawvere's[71] analysis of the definitions of *Mengen* and *Kardinalen* given by Cantor:

---

71 Lawvere 1994.

> By an "aggregate" (*Menge*) we are to understand any collection into a whole [...] *M* of definite and separate objects *m* of our intuition or our thought. These objects are called the "elements" of *M* [...] We will call by the name "power" or "cardinal number" of *M* the general concept which, by means of our active faculty of thought, arises from the aggregate *M* when we make abstraction of the nature of its various elements *m* and of the order in which they are given [...] Since every single element *m*, if we abstract from its nature, becomes a "unit", the cardinal number $\overline{\overline{M}}$ is a definite aggregate composed of units (*lauter Einsen*), and this number has existence in our mind as an intellectual image or projection of the given aggregate M.[72]

Zermelo eliminated this definition of *Kardinalen* in his edition of Cantor's complete works, finding it contradictory. For him, a *Menge* corresponded more or less to our current vision of the whole, an "aggregate" of elements whose only property lies in its own distinguishability, so that any subsequent process of abstraction would inevitably lead to the collapse of these elements. Denote by $\mathcal{U}_i$ Zermelo's interpretation of the sign of *All menges*, which corresponds to a category that we can identify without much loss of generality with the category of sets. In another direction, Cantor seems to have had a much richer vision of *menge*, more in tune with his previous work on number theory and analysis. Thus, from the Cantorian point of view, these entities can continue to undergo fruitful processes of abstraction. Let us denote by $\mathcal{U}$ the Cantorian interpretant of *All menges*, a notion that contemporary mathematics has been recovering little by little, and that we can think of as a cohesive (∞)-topos. Basically cohesion consists of the tendency of the elements of an object to clump together or attract each other. Hence the misunderstanding between these two mathematicians can be explained in terms of different "degrees of determination" of the sign in question.

We can analyze this situation in terms of the triple adjoint theory that we have seen as follows:

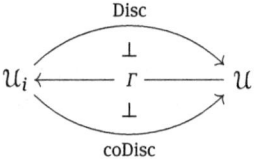

Γ works as an indeterminacy functor: for $M \in \mathcal{U}$, the more determined interpretant of *Menge*, its image $\Gamma M \in \mathcal{U}_i$ is an usual set. Since Γ is a forgetful functor, we can assume without much loss of generality that it possesses a left adjoint, the free functor Disc. The back and forth between them determines the flat modal-

---

[72] Cantor 1915, pp. 85–86.

ity ♭ := Disc ∘ Γ, which *frees the signs from their sedimentary semantic load*[73] and then replicates their image in 𝒰. Therefore, ♭M filters M of any cohesion it may have, but at the same time it opens the opportunity to unite its points in all possible ways. If Γ has a right adjoint coDisc, we can also define the sharp modality ♯ := coDisc ∘ Γ. The latter, which is in opposition to the first (♭ ⊣ ♯), represents the perfect cohesion of all the points of M to the point of making them indistinguishable. ♯M is the necessary state to which ♭M tends if we were to submit its points to an indefinite process of cohesion. The unity and the county of the adjunctions define natural maps ♭M → M → ♯M. Any point on this interval (that is, a factorization of ♭M → ♯M) respects the flow of the cohesion. Zermelo's interpretation is blind to this richness since Γ reduces all this flow to the point ΓM:

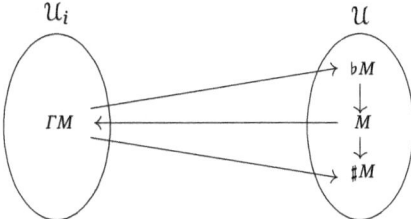

The transit of modalities when unfolding this point in an interval reveals the cohesion, which can be thought of as a (somewhat crude) form of continuity.

The determination process consists of delimiting the entity with increasingly small intervals or, analogously, sifting it with increasingly fine networks that better reveal its essence. The idempotency of modalities implies that at each step we need new adjunctions of (co)monads. Thus, the process of determination consists in the evolution of these adjunctions on a continuum[74] (*Figure 1.3*). In category theory there are several ways in which these adjunctions can evolve. The most immediate is to consider longer adjoint strings. Thus, one of four functors

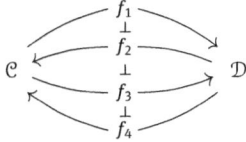

induces a chain "monad adjoint left to comonad, adjoint left to monad",

---

[73] Zalamea 2012a p. 63.
[74] Zalamea 2012a, p. 25.

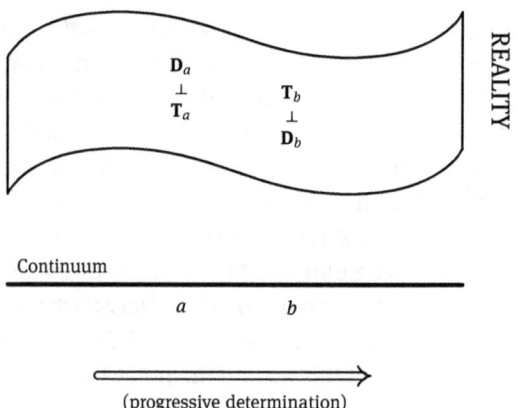

Figure 1.3: (co)Monadical determination in the indeterminate "fibers" of the continuum

$$\mathcal{C} \underset{f_4 f_3}{\overset{f_2 f_1}{\rightleftarrows}} \mathcal{C}$$
$$\dashv f_2 f_3 \dashv$$

and another chain "comonad adjoint left to monad, adjoint left to comonad",

$$\mathcal{D} \underset{f_3 f_4}{\overset{f_1 f_2}{\rightleftarrows}} \mathcal{D}$$
$$\dashv f_3 f_2 \dashv$$

Proposition 1.10. A chain of adjunctions of (co)monads (necessarily interleaved) ♣ ⊣ △ ⊣ ♠ is equivalent to an adjoint pair in the 2-category whose morphisms are adjoint pairs in the original 2-category, hence an adjunction of adjunctions (♣ ⊣ △) ⊣ (△ ⊣ ♠), which can be denoted:

$$\begin{array}{ccc} \clubsuit & \dashv & \triangle \\ \dashv & & \dashv \\ \triangle & \dashv & \spadesuit \end{array}$$

This process can be continued to longer attachment chains. If we think of each adjoint of (co) monads as a sign, this procedure can be thought of as analogous to the Peircean idea that a sign can be determined by putting it in relation to another sign.[75]

We also have the *Aufhebung*, which was developed by Lawvere[76] for adjunctions of the form $\mathbf{D} \dashv \mathbf{T}$, thinking of them as mathematical models of the dialectics of Hegel's *Wissenschaft der Logik* and Mao Tse Tung's *On contradiction*. The initial observation is, as Definition 1.14 suggests, that the two adjunctions $\mathbf{T} \dashv \mathbf{D}$ and $\mathbf{D} \dashv \mathbf{T}$ generate subcategories. These have special characteristics depending on which one we are considering and are enough to characterize it.

Proposition 1.11. An adjoint of the form
1. $\mathbf{T} \dashv \mathbf{D}$ defines and is defined by a bireflective subcategory (i.e., both reflective and coreflective).
2. $\mathbf{D} \dashv \mathbf{T}$ defines and is defined by an essential subcategory.

In the case $\mathbf{D} \dashv \mathbf{T}$, if $\mathcal{U}$ has some good properties (complete and cocomplete, locally small, with either a strong generator or a strong cogenerator[77]), then its essential subcategories form a complete lattice $\mathcal{L}$. Thus, an essential subcategory $\mathcal{U}_i \in \mathcal{L}$ defines and is defined by a pair $\mathbf{D}_i \dashv \mathbf{T}_i$ which in turn defines, and is defined by, a pair of subcategories of $\mathcal{U}$ (the two embeddings of $\mathcal{U}_i$ by the functors r, l of Definition 1.14).

Definition 1.15.
- $Sh_i(\mathcal{U})$, the *i-sheaves*, is the category of algebras associated with $\mathbf{T}_i$. An $X \in \mathcal{U}$ will belong to $Sh_i(\mathcal{U})$ iff $\mathbf{T}_i X \simeq X$. The terminology arises from the Lawvere–Tierney operator.
- $Sk_i(\mathcal{U})$, the *i-skeleta*, is the category of coalgebras associated with $\mathbf{D}_i$. An $X \in \mathcal{U}$ will belong to $Sk_i(\mathcal{U})$ iff $\mathbf{D}_i X \simeq X$. The terminology arises from the skeleton associated with a simplicial set.

Definition 1.16.
1. Given $i, j \in \mathcal{L}$, we can express the usual order $i \leq j$ of $\mathcal{L}$ in several ways. First of all, this can be asserted saying that $Sh_i(\mathcal{U}) \subseteq Sh_j(\mathcal{U})$ and $Sk_i(\mathcal{U}) \subseteq Sk_j(\mathcal{U})$ ($\subseteq$ is the usual subcategory order). By Definition 1.15, this is equivalent to the modal formulas $\mathbf{T}_j \mathbf{T}_i = \mathbf{T}_i$ and $\mathbf{D}_j \mathbf{D}_i = \mathbf{D}_i$.
2. We say that j resolves (resp. coresolves) the opposite of level i, and we will write $i \preceq j$ (resp. $i \overline{\preceq} j$), if $i \leq j$ and also $Sk_i(\mathcal{U}) \leq Sh_j(\mathcal{U})$ (resp. $Sh_i(\mathcal{U}) \leq Sk_j(\mathcal{U})$). This is equivalent to the modal formula $\mathbf{T}_j \mathbf{D}_i = \mathbf{D}_i$ (resp. $\mathbf{D}_j \mathbf{T}_i = \mathbf{T}_i$).

---

75 [**1931–58**] CP 5.505.
76 Lawvere 1989, with later developments in Lawvere 1994, Lawvere 1996, among others.
77 Kelly and Lawvere 1988.

3. A i' (resp. i') is called the *Aufhebung* (resp. co*Aufhebung*) of level i, which we will denote $i \triangleleft i'$ (resp. $i \overline{\triangleleft} i'$), iff it is a minimal level which resolves (resp. coresolves) the opposites of level i, i.e. iff $i \preceq i'$ ($i \overline{\preceq} i'$) and for any k with $i \preceq k$ ($\overline{\preceq}$) then $i' \leq k$ ($i' \leq k$).

If $\sqsubset \in \{\leq, \preceq, \overline{\preceq}, \triangleleft, \overline{\triangleleft}\}$, we will graphically represent any of the above situations as follows:

$$\begin{array}{ccc} \mathbf{D}_i & \sqsubset & \mathbf{D}_j \\ \dashv & & \dashv \\ \mathbf{T}_i & \sqsubset & \mathbf{T}_j \end{array}$$

To explain how this process works in pragmatic terms, let us remember that every continuum has the property of reflexivity, that is, that each of its parts has the same properties of the whole.[78] We can think of $\mathcal{U}$ as a kind of continuum under construction, which uses the desired reflexivity to promote its own evolution (*Figure 1.4*): given a level of determination $\mathcal{U}_i$ *partially undetermining the determined* via functors $\Gamma_{(\_)}$, then we see the transit of modalities generated by the dialectic initiated by this functor in $\mathcal{U}$ and finally, by *Aufhebung, partially determining the undetermined* finding a new level $\mathcal{U}_j$ that reflects this transit.[79] In adjunctions of the form $\mathbf{T} \dashv \mathbf{D}$, since there is only a single bireflective subcategory involved, no sublimation is done and we only use the usual order of this class of subcategories. Graphically

$$\begin{array}{ccc} \mathbf{T}_i & \subseteq & \mathbf{T}_j \\ \dashv & & \dashv \\ \mathbf{D}_i & \subseteq & \mathbf{D}_j \end{array}$$

A systematic study of all these possibilities of determination could bring us closer to the continuum on which, as shown in *Figure 1.3*, (co)monadical adjunctions evolve. The ultimate end of this process should be the determinate, where "no latitude of interpretation"[80] is possible. We can represent this level by the adjoint $id \dashv id$, whose associated subcategory is the same $\mathcal{U}$ and where for all $c \in Ob(\mathcal{U})$ its

---

[78] [1982–2009] W 3.103, "The Conception of Time Essential in Logic". For a study of this property see section 1.4 of Zalamea 2012a, or Vargas's *Chapter 2* in this volume.
[79] For a discussion of this double process of *saturation and freeness*, see Zalamea 2012a, p. 63.
[80] [1931–58] CP 5.447.

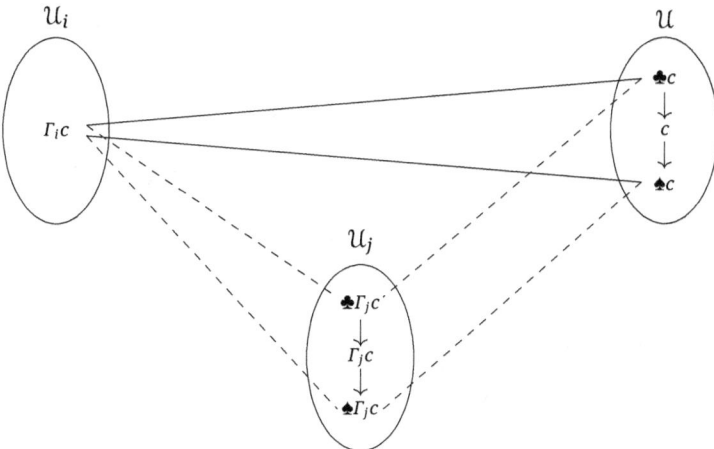

Figure 1.4: *Aufhebung:* reflection of the transit of modalities

interval of determination is reduced to the same object. This shows us that it is an ideal state, which nevertheless serves as a guide for the development of the semiotic process.

Example 1.4. In his essay *New Elements*,[81] Peirce presented his vision of the origin and evolution of reality through a determining process. Recent work by Lawvere and his school shows how physical reality arises precisely in this way, through a process of "categorical determination".[82] The starting point of everything is, according to Peirce, an utter indetermination, where there was nothing in a determined sense, no reaction, no quality, no matter, no consciousness, no space or time, nothing at all.[83] We can represent this state by a singleton, $\{*\}$, in which everything is perfectly fused. This indeterminacy of the absolute beginning is a symbol, which as such produces an infinite series of interpretants. The final interpretant of this symbol is "the *perfect Truth*, the absolute Truth, and as such (at least, we may use this language) would be the very Universe".[84] Let's denote it by $\mathcal{U}$ and assume it to be an (∞)-topos, that is, a rich enough universe to do things. If $\{*\}$ is an essential subtopos of $\mathcal{U}$, then it can be characterized by an adjoint of

---

[81] Included in [1999–98] EP 1.300–324.
[82] For a unified view, see for example Schreiber 2016. For a study in terms of Hegelian philosophy nLab authors 2021b.
[83] [1999–98] EP 2.322.
[84] [1999–98] EP 2.304.

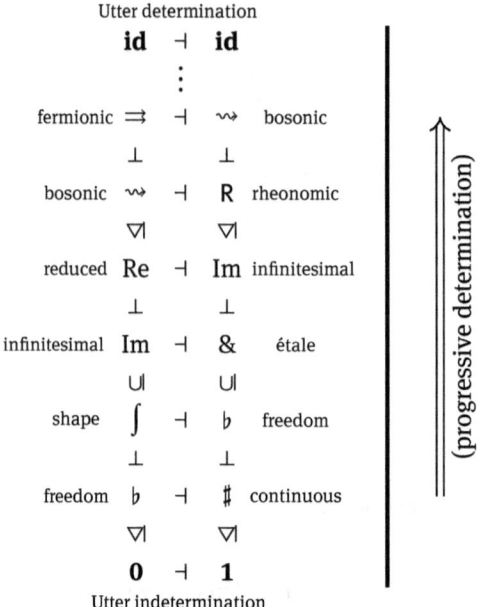

Figure 1.5: Progressive determination of reality (for a discussion, see below)

the form **D ⊣ T** (Proposition 1.11). Let us denote it suggestively by **0 ⊣ 1** and the triple it gives rise to by

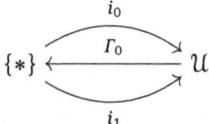

Here $\Gamma_0 : \mathcal{U} \to \{*\}$ is again the indetermination functor, which "frees the signs from their sedimentary semantic load"[85] and submerges them in an utter indeterminacy. By properties of adjointness, $\mathbf{0} := i_0(*)$ is an initial object of $\mathcal{U}$. The comonad $(\mathbf{0}, !_{(\_)}, id_0)$ given by this pendulum oscillation, sends all objects from $\mathcal{U}$ to $\mathbf{0}$, thus defining a replica of indeterminacy in reality. The monad $(\mathbf{1}, !_{(\_)}, id_1)$, which arises from the functor i_1, right adjoint of Γ, defines another copy of the indetermination in $\mathcal{U}$, and analogously gives origin to a final object **1**. For any $t \in Ob(\mathcal{U})$, this attachment does not give us any information that determines it

---

[85] Zalamea 2012a, p. 63.

within the continuum $\mathcal{U}$, except for the fact that we can limit it in the interval $\mathbf{0} \to t \to \mathbf{1}$. We can think of $\mathbf{0}$ as a state of absolute possibility, from which everything is feasible, and $\mathbf{1}$ as absolute necessity, inescapable. The existence of this interval is an invitation to future measurements, that is, to future determinations.

According to *Figure 1.4*, we can get out of this state of utter indetermination by means of an abductive jump (*Aufhebung*) to a level that reflects the transit determined by $\mathbf{0} \dashv \mathbf{1}$.

Proposition 1.12. The Aufhebung $\mathcal{U}_j$ of $\mathbf{0} \dashv \mathbf{1}$ has the property that $\mathcal{U}_{\neg\neg} \leq \mathcal{U}_j$ and coincides with $\mathcal{U}_{\neg\neg}$ in case the latter is an essential subtopos.[86]

We can assume without much loss of generality that $\mathcal{U}_{\neg\neg}$ is the *Aufhebung* of $\mathbf{0} \dashv \mathbf{1}$. This is the smallest essential subtopos where both $\mathbf{0}$ and $\mathbf{1}$ are sheaves (in the sense of Definition 1.15), so in it we can speak of a transit between utter possibilty, actuality and utter necessity, just like we did in $\mathcal{U}$. This new subtopos is determined by the adjunction $\flat \dashv \sharp$, which, as we saw in Example 1.3, are the modalities that allow us to speak of discreteness and cohesion, respectively. On the other hand, let's remember that $\mathcal{U}_{\neg\neg}$ is a Boolean topos, that is, its internal logic is the standard logic. In this way, one confirms that the first interesting level of determination after utter indetermination is the classical universe, rather artificial, quite far from the true reality $\mathcal{U}$.

We can continue the determination process by requiring that the flat modality have a left adjoint, $\int \dashv \flat$, known as shape modality. It sends an object X to its geometric realization (or equivalently, its fundamental ∞-groupoid), so we can think of $\int X$ as the shape of X. Its origin can be seen more clearly from the indetermination functor adjunctions

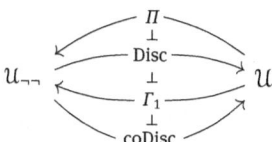

Since $\flat = \text{Disc} \circ \Gamma_1$ then $\int = \text{Disc} \circ \Pi$. Thus, the functor Γ_1 works as a Foucault pendulum that in each oscillation moves it trajectory, with each back and forth providing new information about reality. The shape of an object is the first impression we have of it, something present and immediate that comes to us when we see it, and therefore a *first*. This first feature is technically revealed, in the appearance of Disc, which as we saw is a *free* functor. The appearance of $\int$, forces $\flat$,

---

86 Lawvere and Menni 2015.

which was a *first*, to move its position, now playing the role of a mediator between this new modality and ♯ ($\int \dashv \flat \dashv \sharp$), and thus becoming a *third*.

The triple adjunction $\int \dashv \flat \dashv \sharp$ reveals interesting topological properties of truth $\mathcal{U}$, although for now it remains a very imperfect continuum. To continue with its determination, we sublimate the level $\mathcal{U}_{\neg\neg}$ to a new one $\mathcal{U}_{red}$ that reflects the transit that we have already determined.

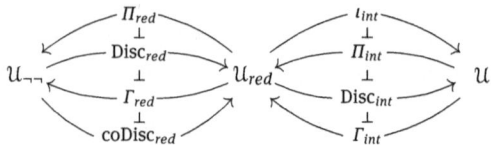

The new indetermination functor Γ_{int} generates a dialogue via adjunctions between $\mathcal{U}_{red}$ and $\mathcal{U}$ thanks to which, in each coming and going, new properties are revealed of the latter through the transit between modalities. The new modalities, the infinitesimal flat $\& := \text{Disc}_{int} \circ \Gamma_{int}$, infinitesimal shape $\text{Im} := \text{Disc}_{int} \circ \Pi_{int}$ and the reduction $\text{Re} := \iota_{int} \circ \Pi_{int}$, are infinitesimal analogs of those we studied previously, and bestow upon $\mathcal{U}$ sort of properties of differential geometry.

To finish this part of our discussion, we sublimate the level $\mathcal{U}_{red}$ to a new $\mathcal{U}_{bos}$ that reflects the new transit between differential modalities determined previously

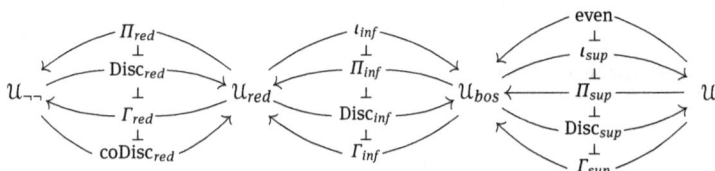

This *Aufhebung* also gives rise to an indetermination functor Γ_{sup}, which in its pendular coming and going, generates the modalities $\text{R} := \text{Disc}_{sup} \circ \Pi_{sup}$, bosonic $\rightsquigarrow := \iota_{sup} \circ \Pi_{sup}$ and fermionic $\Rightarrow := \iota_{sup} \circ$ even. These provide the reality of a supergeometry, generalization of differential and algebraic geometry, necessary to do field physics.

The process could be continued in the same way (looking at each step for a part that adequately reflects the transit of modalities of the whole), but at this point it seems rich enough to model some of the latest advances in contemporary geometry and physics. This allows us to appreciate the enormous richness of the Peircean continuum and the complexity of the seemingly innocent property of reflexivity. *Ad infinitum*, could we arrive at a model of the Peircean continuum in this way? In *Figure 1.6* we see how, as the evolution of adjoint (co)monads progresses (*Figure 1.5*), the different modal transit intervals better delimit the entity.

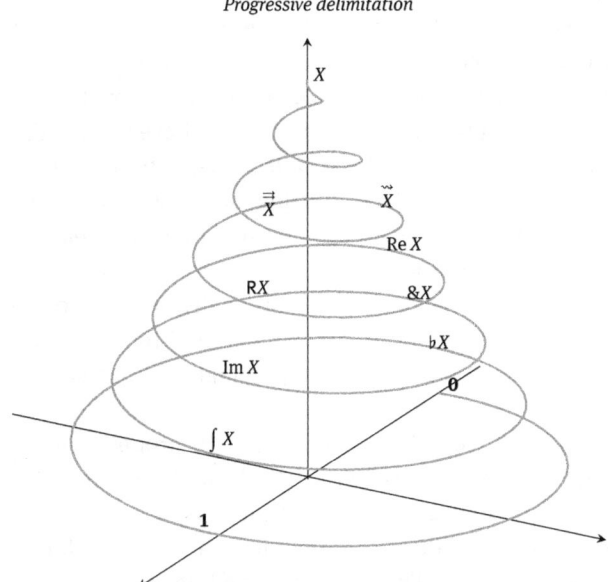

Figure 1.6: Progressive delimitation of the being

1.7 Further Venues

In this chapter we wanted to show how the constructions that arise in higher category theory, such as higher groupoids, sheaves and monads, are closely related to the original ideas of Peirce, and can express them better than some of the incipient mathematical structures of the beginning of 20th century. However, higher category category is still in its infancy, so there are many lines of research ahead. To begin with, which of the different proposals best suits Peircean thought? Here we have used the model based on simplicial sets (quasi-categories) as it is the best known and most developed, but it would be necessary to analyze the virtues and disadvantages of other proposals. For example, globular sets are in better correspondence with our intuition of what n-morphisms should look like, but present difficulties in modeling homotopy theory n-types.

Here we have focused mainly on the semiotic flow and the main instrument to apprehend it, the Pragmatic Maxim. However, there are two other fundamental columns of Peircean architecture of which we made little mention in the text, and of which we would like to say a few words. First of all, we have Existential Graphs. Already the innovative work of Brady and Timble proposed categorical semantics for the graphs: monoidal categories with a suitable strong functor ¬ for the Alpha part[87] and string diagrams satisfying Beck–Chevalley for the Beta part.[88] This is precisely the context of the constructions that we have used in this work: monoidal categories of various types for the Rosetta stone,[89] strong monads on monoidal categories for computational effects,[90] string diagrams and symmetric monoidal categories for descent theory[91] Along the same lines, Corfield has conjectured that higher modal category theory is a good approximation for the Gamma part of the graphs.[92] Does higher category theory really offer a good semantics for Existential Graphs? If so, given our approximation of the Pragmatic Maxim, following this path could we obtain a proof of pragmatism in Existential Graphs?

On another vein, we already saw on p. 10 that the Kan lift property can be thought of as a form of abduction. In an entry in nLab[93], the authors propose to study deductive, inductive and abductive reasonings by means of composition, extension and lifting of morphisms. To date, there seems to be no further information on this conjecture. However, if it is valid, it opens up an interesting number of lines of investigation. For example, as we saw in our Remark p. 30, in homotopy type theory (the internal language of ∞-topos) identity between terms is conceived as a path continuously varying over a type. Path elimination is the procedure that allows us to show the validity of any identity property in the future based on its validity in the present.[94] Since it corresponds to a "path-lifting" property in (higher) category semantics, can this kind of reasoning correspond to a form of abduction? This seems to agree with Peirce's view: "An Abduction is a method of forming a general prediction without any positive assurance that it will succeed either in the special case or usually, its justification being that it is the only possible hope of regulating our future conduct rationally, and that Induction from past

---

[87] Brady and Trimble 2000.
[88] Brady and Trimble n.d.
[89] Baez and Stay 2011
[90] Uustalu and Vene 2008
[91] Obradovic 2016.
[92] Corfield 2020, pp. 138.
[93] nLab authors 2021a
[94] For a technical presentation, see The Univalent Foundations Program 2013, p. 50.

experience gives us strong encouragement to hope that it will be successful in the future".[95] Given the great proliferation of lifting problems in higher category and homotopy theory, do they constitute a natural setting for the study of abduction?

Here we have focused mainly on putting pragmatism in dialogue with the theory of categories. However, both are included in a much more general science of systems and processes, so that many other "mixtures" (in the Lautmanian sense) are possible. Already the connection between pragmatism and computational theory has revealed a number of interesting works in the new fields of computational pragmatics (the computational study of the relationship between signs and contexts[96]) and pragmatic computation (the use of pragmatic techniques and ideas to solve computational problems; for example, employ abductive reasoning in artificial intelligence[97] or a semiotic approach to design user interface languages.[98]) The mixtures between pragmatism and physics have also resulted in a great deal of work. However, we think that the application of new structures that have emerged to model the physical universe, such as higher modal category theory (which we review in Example 1.4) and Wolfram's models[99] still have much to say in Peircean mathematical studies.

---

95 [1931–58] CP 2.270.
96 For an introduction see Bunt and Black 2000.
97 Gabbay and Kruse 2000.
98 de Souza 2005.
99 Wolfram 2020.

Francisco Vargas
2 A Full Model for Peirce's Continuum

Abstract: This chapter aims to provide a set-theoretic reconstruction—a model—of the Peircean ideas on the continuum (non constituted by ultimate elements, self-mirroring, potential and beyond any transfinite cardinal). Going further, different paths of mathematical development are explored allowing ramifications through infinitesimal approaches in Algebra, in Geometry, and finally, in Analysis.

Keywords: continuum; synechism; infinitesimals; mereology; microstraightness

An Alternative Continuum Prospect

The theme of continuity is pervasive along the development of the philosophy of Charles Sanders Peirce. Even so, it is in the expression of his mature thought that the topic occupies the center of his system. He comes then to label his philosophy as "Synechism", which he defines in 1893 as "the doctrine that continuity rules the whole domain of experience".[1] Continuity turns out to be "the master key which adepts tell us unlocks the arcana of philosophy".[2] It fits Peirce's triadic thought where difference, opposition and Secondness are transcended through mediation, as deployed in a process of infinitary semiosis. All realms of phenomena disclose variation by infinitesimal degrees.[3] This leads the demand to withdraw our reliance in well defined boundaries and our atomistic belief in ultimate constituent components.

Continuity was approached by the author over and over again throughout his life, leaving us as a result a multiplicity of ideas that take up a whole philosophical tradition that goes back to the Greeks, but at the same time is inspired by and makes use of modern mathematical tools of topology, logic and the theory of transfinite sets. Peirce integrates these ideas by rethinking them in a swing that

---

[1] MS 946, p. 5.
[2] [1931–58] CP 1.163 (Summer 1893).
[3] This is valid not only for physical, but for psychical processes as well, as established by his "Law of Mind", *cfr.* [1892].

Francisco Vargas, Francisco Vargas (1977) is Professor at Universidad Pedagógica y Tecnológica de Colombia.

oscillates between the germinal state of the birth of new concepts and their quasi-axiomatic technical refinement.[4] His thoughts on continuity are aware of other attempts to approach the subject, such as Aristotle's or Cantor's. It integrates aspects of both, but goes well beyond them in highly original ways. Nevertheless, these conceptions have been considered by some scholars nothing more than "a castle in the air".[5] What Peirce says in fact deviates not only from the standard construction of the continuum as the set of real numbers, but also from the better known 20th century alternatives such as the intuitionistic conceptions[6] or Robinson's hyperreals.[7]

In his 1898 *Cambridge Lectures*[8], Peirce vindicated the consistency of his conceptions on what he calls "true continuity", emphasizing that his "logic of relatives" is suitable for proving it: "that apparatus not only absolutely refuses to pronounce this self-contradictory but it demonstrates, on the contrary, that it is not so". This proposal remained unaccomplished or, at least, not completely elaborated in a definitive way. Even so, clear advances towards actual mathematical constructions are present in Peirce's manuscripts themselves (see *Subsection 2.1.3* below).

The main aim of this *Chapter 2* is to provide a systematic presentation to the construction for a continuum as introduced by Vargas[9] with further developments in different directions than previously provided. *Section 2.1* is a reminder of Peirce's main ideas on the topic and how he himself proposed explicitly a construction in the direction here offered. *Section 2.2* introduces the basic definitions and properties of the model presented and shows how they fit the main Peircean requirements for a continuum. *Section 2.3* explores the purely mereological prop-

---

[4] The different conceptions developed by Peirce may seem overly confusing and even contradictory if not approached from a chronological perspective. Let us remind Havenel 2008, with his subdivision of Peirce's views into five major periods: Anti-nominalistic Period (1868–1884), Cantorian Period (1884–1892), Infinitesimal Period (1892–1897), Supermultitudinous Period (1897–1907) and Topological Period (1908–1913). Consistently with the use of different scholars (*e.g.* Zalamea 2012a), I will refer in what follows to the Supermultitudinous conception indicated simply as "Peirce's continuum". This use is motivated by the fact that, as remarked in Moore 2015, "it is far more original than those that preceded it, and far more fully developed and widely applied than any of the others".

[5] Murphey 1961, p. 407.

[6] For a recent account, see van Dalen 2009. At first instance, the use of *constructive* tools in intuitionism (such as species, fans, choice sequences) deviates from Peirce, but, as *Chapters 1* and *3* show, a natural intertwining between Peirce and intuitionism happens *a posteriori*.

[7] Robinson 1974, Goldblatt 1998.

[8] [1993].

[9] Vargas 2015 and Vargas and Moore 2021.

erties of the model establishing some basic connections with pointless geometry as developed from Whitehead's ideas. In *Section 2.4* it is shown how the main construction induces the definition of a Kripke model that renders explicit the modal character of the continuum. *Section 2.5* shows how the construction not only may be developed in the direction of the infinitely small, but can be mirrowed also towards the infinitely large. Similarly, in *Section 2.6* it is shown how the construction may be adapted in order to go beyond 1-dimensional continua to higher dimensions. The emphasis will be here on how different directions may be taken for the definition of the plane, something that will be instrumental for developing geometric and analytic ideas. *Section 2.7* goes beyond the order and parthood structure of the model introducing also the algebraic structure inherited in a natural way from the real numbers. *Sections 2.8* and *2.9* elaborate on the development of Geometry and Calculus from an alternative infinitesimal viewpoint. *Section 2.9*, in particular, uses the algebraic definitions of *Section 2.7* which lead to a "micro-straightness" approach to differentiability (close, in this respect to the one of Smooth Infinitesimal Analysis). *Section 2.10* recapitulates the situation and proposes further potential lines of work.

A *unifying, holistic*, trend behind these particular sections, is the study and development of Peirce's continuum in all its *richness* from a multiplicity of "views": size, geometrical locus, modal environment, infinitesimal perspectives, calculability features. Through all these approaches, the *dense plasticity* of Peirce's continuum, perhaps its essential character, shines forcefully.

2.1 Peirce's Riddles on Continuity

Peirce's hypotheses on what he calls "true continuity" manifest themselves more as an evolving (albeit converging) path, aproaching one of the most eluding ideas in philosophy,[10] than as a completely finished system ready for use in philosophy or science.

To start by a negative characterization, it is important to highlight the enormous distance that separates the mature Peircean ideas on the continuum (particularly around the turn of the century) from today's dominant conceptions which identifies it with the real numbers. Indeed, leaving apart the property of density, among the main characteristics that Peirce attributes to the continuum ("inextensibility", "supermultitude", "reflexivity", "potentiality" and "genericity"[11]),

---

10 "Of all conceptions Continuity is by far the most difficult for Philosophy to handle" [1993].
11 See Zalamea 2012a.

none (!) is present in the construction of the reals *à la* Cantor–Dedekind. This construction, nowadays assumed as standard, incarnates for Peirce only an "embryo of continuity", a continuum in a nascent state yet, far from having been deployed in all its potential beyond all "multitudes" (of any transfinite cardinal) and beyond "extensibility" (of the existence of points that determine it). This *Section* recapitulates for the sake of completeness the main features of continuity according to Peirce.[12]

2.1.1 Thinking Back on Points, Parts and Boundaries

Let's start by the fundamental "Aristotelian" feature of the continuum: not being constituted by ultimate parts.

> [...] no collection of points, each distinct from every other, can make up a line, no matter what relation may subsist between them.[13]

Thus, the continuum is *inextensible*, not analytical, and points are only conceivable, if at all, as ideal limits. Points do not have actual existence, or assume actual existence only through the negation of continuity itself, breaking it: "Hence a point or indivisible place really does not exist unless there actually be something there to mark it, which, if there is, interrupts the continuity".[14]

Thinking continuity as point-based leads to problems and contradictions as long-standing as Zeno's paradoxes. Moreover, this assumption is tied to philosophical nominalism, a doctrine opposed to Peirce's epistemic fallibilism.

> The ordinary scientific infallibilist... cannot accept *synechism*,—or the doctrine that all that exists is continuous,—because he is commited to discontinuity in regard to all those things which he fancies he has exactly ascertained, and especially in regard to that part of his knowledge which he fancies he has exactly ascertained to be *certain*.[15]

Peirce's position, of course, is completely opposed to the usual Cantor–Dedekind approach in analysis. The historical process of arithmetization of the field has led to identify the line with the real numbers (thus conceived as points).[16]

---

12 As in the correspondent sections of Vargas 2015 and Vargas and Moore 2021 I will follow the terminology in Zalamea 2012a.
13 [1931–58] CP 4.640.
14 [2010] selection 18.
15 [2010] selection 20.
16 For Peirce views on an arithmetic continuum see *e.g.* [1976] NEM 3.93, 3.127.

Peirce takes the non-punctual character of the continuum further, affirming that on it "every part has parts of the same kind". The continuum mirrors, reflects itself in each of its parts. This *reflexivity* leads to a "uniform ontology", in contrast to the mereological part/whole asymmetry characteristic of the usual point-based approaches.

> A continuum (such as time and space actually are) is defined as something any part of which however small itself has parts of the same kind. Every part of a surface is a surface, and every part of a line is a line. The point of time or space is nothing but the ideal limit towards which we approach indefinitely close without ever reaching it in dividing time or space.[17]

Points are just ideal limits. They are not conceived as the ultimate substantial constituents of a continuum, and, therefore, properties may not be predicable of them:

> [...] it is only as they are connected together into a continuous surface that the points are colored; taken singly, they have no color, and are neither black nor white, none of them.[18]

Vagueness is invoked here, as "absolute exactitude of thought is quite impossible".[19] Extended entities reveal their non-binary (more generally: non discrete) character. This way, boundaries manifest themselves also as transitions. To use the author well-known example in "The Law of Mind", the boundary of the red part and the blue part of a surface "is half red and half blue", and a similar situation happens to the boundary between times: "the present is half past and half to come".[20]

This leads, ultimately, to revise the grounds of logic itself,[21] thinking on a "logic of things" which transcends for instance, the principle of excluded middle:

> [...] we must either say that a continuous line contains no points or we must say that the principle of excluded middle does not hold for these points. The principle of excluded mid-

---

17 [**1993**] 3.103. For Peirce this property is tied to the character of general law of his category of Thirdness: "A perfect continuum belongs to the genus, of a whole all whose parts without any exception whatsoever conform to one general law to which same law conform likewise all the parts of each single part" [**1931–58**] CP 7.535, note 6.
18 [**1931–58**] CP 4.127 (1893).
19 [**2010**] p. 131.
20 [**1931–58**] CP 6.126 (1892).
21 Zalamea 2012a highlights the need to develop a vagueness logic and a neighbourhood logic as local approaches to Peirce's continuum.

dle only applies to an individual (for it is not true that "Any man is wise" nor that "Any man is not wise"). But places being mere possibles without actual existence are not individuals.[22]

2.1.2 Cardinal Transcendence, Modality and Genericity

Peirce was deeply influenced by the emergence of Cantorian set theory and he contributed in different ways to the conceptualization of the infinite.[23] In different points, Peirce's ideas on the topic diverged from what became standard views on it, as for example in regard to his potential or "supermultitudinous" (= larger than any definite infinite cardinal) collections. *Supermultitudinousness* turns out to be a characteristic feature of the continuum, and the concept intertwines with its nonpunctual nature. Going beyond all transfinite cardinals is also a transcendence of the actual existence of points which leads to inextensibility.

A true continuum should go well beyond the property of density (already present in the rational numbers):

> The idea of a general involves the idea of possible variations which no multitude of existent things could exhaust but would leave between any two not merely many possibilities, but possibilities absolutely beyond all multitude.[24]

> [...] any multitude of points whatever are determinable on the line (not, of course, by us, but of their own nature).[25]

Supermultitudinousness leads to indiscernibility, where points are "welded":

> A supermultitudinous collection is so great that its individuals are no longer distinct from one another [...] A supermultitudinous collection, then, is no longer discrete; but it is continuous.[26]

The dichotomy between the actual and potential determination of places and points makes the continuum an essentially *modal* entity:

---

22 [2010] selection 18. Let us remind that going even further Peirce prefigured a radical convulsion of logic: "...individuals would be sunk to a potential being, and would no longer be unconditionally and *per se* there. The discovery of such a state of things would be an earthquake in logic, leveling its whole fabric; and it would be incumbent upon the philosopher who should accept it to begin at the very beginning and build up the elementary rules of reasoning anew" [2010] selection 21.
23 Dauben 1982, and Moore 2010.
24 [1931–58] CP 5.103.
25 [1931–58] CP 3.568.
26 [1976] NEM 3.86–87.

> That which is possible is in so far general and, as general, it ceases to be individual. Hence, remembering that the word 'potential' means indeterminate yet capable of determination in any special case, there may be a potential aggregate of all possibilities that are consistent with certain general conditions; and this may be such that given any collection of distinct individuals whatsoever, out of that potential aggregate there may be actualized a more multitudinous collection than the given collection. Thus, the potential aggregate is, with the strictest exactitude, greater in multitude than any possible multitude of individuals. But being a potential aggregate only, it does not contain any individual at all. It only contains general conditions which permit the determination of individuals.[27]

> We must, therefore, conceive that there are only so many points on the line as have been marked, or otherwise determined, upon it. Those do form a collection; but ever a greater collection remains determinable upon the line. All the determinable points cannot form a collection, since, by the postulate, if they did, the multitude of that collection would not be less than another multitude. The explanation of their not forming a collection is that all the determinable points are not individuals, distinct, each from all the rest.[28]

From the wider perspective of Peirce's metaphysics[29] this modal approach is also connected with "generality"[30]:

> The idea of a general involves the idea of possible variations which no multitude of existent things could exhaust but would leave between any two not merely *many* possibilities, but possibilities absolutely beyond all multitude.[31]

For Peirce, "continuity and generality are two names for the same absence of distinction of individuals".[32] This leads to regularity and uniformity, as usually expected in a continuum[33]:

---

27 [1976] NEM 3.106.
28 (Peirce 1900a:C3, p. 363), [**1931–58**] CP 3.568.
29 In connection with the reality of the category of Thirdness.
30 Zalamea 2012a proposes the equivalent term "genericity", as used in more mathematical and logical contexts.
31 Harvard Lectures on Pragmatism, 1903, [**1931–58**] CP 5.103.
32 [**1931–58**] CP 4.172.
33 This is clearly expressed by the mathematician R. Thom: "the archetypical continuum is a space which possesses a perfect qualitative homogeneity; I would like to say that two "points" are always equivalent by means of a continuous sliding (eventually local) of the space on itself; unfortunately the very notion of a "point" already presupposes a break of spatial homogeneity. [...] The notion of place (Aristotle's τoπóς) could perhaps help to access a rigorous definition." Thom 1992. I will come back to the necessity of going beyond the notion of "point" in *Section 2.3* below from a "pointless" mereological view.

> Continuity is thus a special kind of generality, or conformity to one Idea. More specifically, it is a homogeneity, or generality among all of a certain kind of parts of one whole[34].

> The idea of continuity is the idea of a homogeneity, or sameness, which is a regularity. On the other hand, just as a continuous line is one which affords room for any multitude of points, no matter how great, so all regularity affords scope for any multitude of variant particulars; so that the idea [of] continuity is an extension of the idea of regularity. Regularity implies generality.[35]

2.1.3 Peirce's Proposal for Infinitesimals

In Manuscripts S-14 and 718,[36] we find an explicit way to construct an extension of the "series"[37] of real numbers which allows for the presence of infinitesimals. Peirce describes this construction in order to show how infinitesimals are not inconsistent, despite the dominant conception (after Weierstrass) which ostracized them from mathematics. As is well known, Peirce (and his father, Benjamin Peirce[38]) advocated for the use of infinitesimals in Analysis:

> It is singular that nobody objects to $\sqrt{-1}$ as involving any contradiction, nor, since Cantor, are infinitely great quantities much objected to, but still the antique prejudice against infinitesimally small quantities remains.[39]

For his construction, Peirce's starting point are the real numbers ("the series of numbers, rational and irrational") or the topologically equivalent interval (0, 1).

---

34 [**1931–58**] CP 7.535, note 6.
35 [**1931–58**] CP 7.535.
36 [**1976**] NEM 3.121, 3.125.
37 This is the term used by Peirce, diverging from current usage of "sequences and series" of numbers which presuppose a well-order.
38 B. Peirce 1837.
39 It is worth noticing that in parallel with this kind of reasoning, Peirce's mature conception of infinitesimals is better understood as dealing with "infinitesimal micro-segments" than with "infinitesimal numbers". This is consistent with the non-punctual character of his continuum and the "reflexivity" property: "in a continuous expanse, say a continuous line, there are continuous lines infinitely short. In fact the whole line is made up of such infinitesimal parts" ([**2010**] p. 156). The same argument in defense of using infinitesimals as a manifestation of the natural process of exhausting the potential extensions of numerical systems is also crucial for Gödel: "Arithmetic starts with integers and proceeds by successively enlarging the number system by rational and negative numbers, irrational numbers, etc. But the next quite natural step after the reals, namely the introduction of infinitesimals, has simply been omitted. I think, in coming centuries it will be considered a great oddity in the history of mathematics that the first exact theory of infinitesimals was developed 300 years after the invention of the differential calculus", *cfr.* Robinson 1974.

According to him, we can see each of these numbers not as indicating a point, but a complete set ("series") of points:

> We may use the numbers to designate, each, some one point of a distinct one of those series of points. The number will not show precisely what point is meant; but it will show to which series the point meant belongs.

We can then iterate the process:

> We will now change that feature of the hypothesis, and make them no longer series of points, but series of series; each of the last series being a series of points analogous to the whole series of rational and irrational numbers. We will make another change, so that the objects of every series shall be minuter series.

The process is pursued infinitely:

> The result is, that we have altogether eliminated points. We have a series of series of series, *ad infinitum*. Every part, however closely designated, is still a series and divisible into further series. There are no points in such a line; there is no exact boundary between any parts.

It is clear that Peirce insists, without using his own terminology, in the non-composability or inextensibility of this "line" as well as in the presence of infinitesimals: "there are parts immeasurably smaller than any given part". It is also remarkable that infinitesimals are conceived here algebraically, as numbers, giving rise to different orders of infinitesimals.[40] Nevertheless, he is not explicit here about the supermultitudinousness and the modal character of the construction. The richness of it is not fully developed in the manuscript and, to the best of my knowledge, in any other place of his writings. The construction in next *Section 2.2*, introduced by Vargas,[41] actually fulfills Peirce's proposal and the other Peircean requirements in present-time set theoretic standards.[42]

---

40 "Numbers are equally applicable to these [lengths] also; and then they are algebraically treated as infinitesimals. Again, there will be lengths not measurable by such numbers, nor by limits of series of them. These, when numbers are applied to them, become infinitesimals of the second order" [**1976**] NEM 3.127.
41 Vargas 2015 and Vargas and Moore 2021
42 A similar proposal is also presented in Myrvold's appendix of Myrvold 1995 The use of ordinally indexed sequences of reals and the lexicographic order coincides with the definitions of Vargas 2015 and Vargas and Moore 2021, but the introduction of the *inverse E* relation (Definition 2.4 below) is new (only present in the last references). This turns out to be crucial to obtain formally the full properties of Peirce's continuum, such as reflexivity (Theorem 2.1).

2.2 A Set Theoretic Reconstruction

This section briefly presents the central construction of this chapter, namely the $\mathcal{C}_{Ord}$ model.[43] The construction is carried out using ZFC, the usual axioms of set theory, with the standard notation.[44] Following current usage, too, in what follows (in Definitions 2.3, 2.4 and 2.5, for example) we will deal with proper classes, which should be seen just as the formulae in the (first order) language of set theory that define them, that is, that are satisfied by the members of their respective classes.

The final model will be obtained as a union of "partial models" defined for each ordinal. So, for each $\alpha \in Ord$, the class of all ordinals, we define the "continuum of order α". Peirce would call this a "pseudo-continuum", since supermultitudinousness fails at each particular step. Each of these models can be seen, simply, as the set of real sequences of length α:

Definition 2.1. For $\alpha \in Ord \smallsetminus \{0\}$ let $\langle \mathcal{C}_\alpha, <_\alpha \rangle$ be defined as the set of α-real sequences with the lexicographical order.

We can understand this definition as a process in which we start with $\mathcal{C}_1$ which essentially can be seen as $\mathbb{R}$ (each sequence of length 1 can be identified with the corresponding real, say r_0). In the next step we take the 2-sequences or ordered

---

[43] As discussed in Vargas and Moore 2021, we can see partially reflected some aspects of Peirce's conceptions in previously developed models. In *Section 2.3* we will see some coincidences with the mereological approaches derived from Whitehead. We see analogies also with the intuitionistic approaches (*e.g.* in the potential character of the continuum and the idea that its existence is prior to that of the "points" eventually determinable on it). Nevertheless, given the non-constructive nature of infinitary mathematics "supermultitudinousness" is in principle unapproachable from this perspective. Closer to Peirce's definitions are the comments by Ketner and Putnam regarding the use of Robinson's hyperreals to capture some aspects of them, see C. S. Peirce 1992. This accounts for the existence of infinitesimals and the possibility of having richer models (cardinally speaking) than the real numbers. Other properties, specially inextensibility and reflexivity are not reached through this approach, leading to suggestions such as iterated versions of non-standard models which could be used as fibers on a sheaf, see Martín 2000.

Ehrlich 2010, provides a supermultitudinous model $\langle No_P, < \rangle$ that is nevertheless based on the collection of a class of points, in this case represented by surreal numbers. Connections with smooth infinitesimal analysis (SIA) and Peirce's ideas on infinitesimals are highlighted in Bell 1998, and suggested by Peirce scholars (Zalamea 2012a, Havenel 2008) but no specific construction has so far been given along these lines in order to reflect all aspects of Peirce's continuum. Parallels with SIA in developing Calculus are discussed below in *Section 2.9*.

For a general discussion on other directions see Zalamea 2012a. Other connections will be explored here in the forthcoming sections.

[44] See Kunen 1980 or Jech 1997.

pairs $(r_0; r_1)$. The order is defined in such a way that we can visualize this step as taking each of the points (or what could be thought of at a first glance as a "point") of $\mathcal{C}_1$ and amplifying it in a whole copy of $\mathbb{R}$ (see *Figure 2.1*).

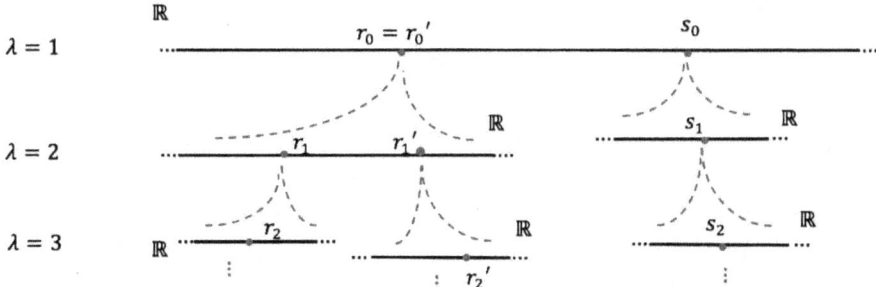

Figure 2.1: We see here a schematic view of how the first three levels of the construction are structured. We have the real line for $\lambda = 1$ and each real number "unfolds" or "explodes" in a multiplicity of other points: a copy of $\mathbb{R}$. The process continues indefinitely over all the ordinals. We can consider in the image elements of $\mathcal{C}_3$ like $r = (r_0; r_1; r_2)$, $r' = (r'_0; r'_1; r'_2)$ and $s = (s_0; s_1; s_2)$. According to the lexicographical order of Definition 2.1, clearly $r < s$. Also, $r < r'$ because, even if the two sequences coincide at the level of $\lambda = 1$, they differ at the following level and $r_1 < r'_1$. We can also visualize the parthood relation: for instance, $(s_0; s_1; s_2)E(s_0; s_1)E(s_0)$.

Each "point" is such only in appearance, and we can repeat this process also at the next level. Iterating through all ordinals, we see that we can "enlarge" each apparent point as much as we want, according to the metaphor of a magnifying glass or microscope used in infinitesimal analysis. The definitions that follow are also inspired by this image. This coincides also with Peirce's proposal summarized in *Subsection 2.1.3*.

We can introduce also in a straightforward way the cumulative models "up to" some stage α:

Definition 2.2. Let $\langle \mathcal{C}_{<\alpha}, <_{<\alpha} \rangle$ be defined as
- $\mathcal{C}_{<\alpha} = \bigcup_{0 < \lambda < \alpha} \mathcal{C}_\lambda$
- For any $x, y \in \mathcal{C}_{<\alpha}$ we define $x <_{<\alpha} y$ iff, for $\lambda = min(dom(x), dom(y))$, $x \upharpoonright \lambda <_\lambda y \upharpoonright \lambda$.

We define also $\langle \mathcal{C}_{\leq\alpha}, <_{\leq\alpha} \rangle$ in the obvious way.[45]

---

[45] In order to avoid cumbersome notation, subscripts will be usually omitted in $<_{<\alpha}$ and $<_{\leq\alpha}$.

Next, for the definition of the class-continuum, we take the union over the whole hierarchy and define the order in a manner consistent with the previous levels:

Definition 2.3. Let $\langle \mathcal{C}_{Ord}, <_{Ord} \rangle$ be defined as
- $\mathcal{C}_{Ord} = \bigcup_{\alpha \in Ord \setminus \{0\}} \mathcal{C}_\alpha$
- For any $x, y \in \mathcal{C}_{Ord}$ we define $x <_{Ord} y$ iff, for $\alpha = min(dom(x), dom(y))$, $x \restriction \alpha <_\alpha y \restriction \alpha$.

Elements of $\mathcal{C}_{Ord}$ can be seen as "places" or "monads" that indicate a location in the straight line, but which are not "points" in the analytical-atomistic sense. As a consequence of this $<_{Ord}$ is not a total order, since it is possible to have "nested" loci which are not comparable according to this order. Points are ideal limits that we would reach if we could specify a place indefinitely. This would correspond to a proper class of loci embedded in $\mathcal{C}_{Ord}$. These are not reachable objects in $\mathcal{C}_{Ord}$, and not even in V, the universe of all sets.[46]

Since all elements of $\mathcal{C}_{Ord}$ considered as places have inner loci within them, it is useful to indicate this by a relation of mereological membership which we call E (to distinguish it from the set theoretical $\in$) defined as follows:

Definition 2.4. For $x, y \in \mathcal{C}_{Ord}$ we have that xEy iff $dom(y) < dom(x)$ and $x \restriction dom(y) = y$.

xEy indicates, in other words, that "x is a member of", or better, "x lies in" the place represented by y. It is clear that in this case x contains more information than y, because it specifies with a larger degree of accuracy a place on the line.[47]

Each $x \in \mathcal{C}_{Ord}$ can be seen from two different points of view: it can be seen as an *element* of this class according to the set-theoretic membership relation, but it can also be seen as a class based on the *reversed* viewpoint of relation E. From this last perspective, seen as a class, it mirrors the structure of the whole model $\mathcal{C}_{Ord}$. To highlight this we introduce formally the concept of "monad".

Definition 2.5. For $x \in \mathcal{C}_{Ord}$ let $\mathcal{M}_x$, the monad associated to x, be defined as the class of elements $y \in \mathcal{C}_{Ord}$ such that yEx.

---

[46] In Vargas and Moore 2021 the "final continuum" constituted by these "points" is indicated as $\mathcal{C}^*$. It lacks sense in ZFC, since it would be a proper class whose "elements" would themselves be proper classes. The "elements" of this class can be imagined as limits of approximations of elements in $\mathcal{C}_{Ord}$, that is, proper class-limits of set-approximations.

[47] As remarked in Vargas and Moore 2021 the E relation is transitive and non-well-founded. We have not only ω-descendent chains, but α-descendent chains for any ordinal α, and even Ord-descendent chains.

We can now see how $\mathcal{C}_{Ord}$ satisfies the Peircean reflexivity property:[48]

Theorem 2.1. For any $x \in \mathcal{C}_{Ord}$

$$\langle \mathcal{M}_x, <_{Ord}, E \rangle \simeq \langle \mathcal{C}_{Ord}, <_{Ord}, E \rangle.$$

Proof. Let $\alpha = dom(x)$. The isomorphism follows because any sequence of ordinals cofinal in Ord is order-isomorphic to this class. So we can perform the same construction as in Definitions 2.1 and 2.3. More formally, $\mathcal{M}_x$ is the class of real-set-sequences whose initial α-subsequence is precisely x. Any sequence $y \in \mathcal{M}_x$ can be written as a concatenation $y = x\!\frown\! y'$. The isomorphism is defined by assigning $y \mapsto y'$. □

This result enables us also to establish, by transitivity, an isomorphism between any two given monads, a first form of homogeneity (see *Subsection 2.1.2* on this aspect of genericity):

Corollary 2.1. For any $x, y \in \mathcal{C}_{Ord}$

$$\langle \mathcal{M}_x, <_{Ord}, E \rangle \simeq \langle \mathcal{M}_y, <_{Ord}, E \rangle.$$

On the other hand, we can also consider homogeneity in the sense of performing a rigid translation establishing an isomorphism of the space into itself. As in $\mathbb{R}$ we can slide $\mathcal{C}_{Ord}$ into itself, even if we don't have any "points". If we move one place to another of the same degree in the hierarchy, we can also move the entire space together with it without altering its global structure.

In the following definition we will use the natural notion of sum between sequences of the same length in $\mathcal{C}_{Ord}$, defined componentwise. Similarly, we can define the opposite $-x$ of a sequence x by taking the sequence of opposite real numbers. We indicate with $x\!\frown\! y$ the concatenation of the sequences x and y. Finally, the notation $(\bar{0})_{<\alpha}$ will indicate the constant sequence of 0's of length α.

---

[48] Proper classes are not indispensable in order to obtain reflexivity. Similar results hold, with the same proof, for $\mathcal{C}_{<\alpha}$ whenever α is a cardinal. However, we take as our model the whole of $\mathcal{C}_{Ord}$ because, it satisfies the property of supermultitudinousness, and more faithfully reflects the modal dichotomy between potential and actual. One aspect of this is present in the opposition between "sets" and "proper classes", which distinguishes the level of objects achievable by the theory (in this case ZFC), from the metatheoretic level in which we can refer to the formulas of this theory (which *are* by definition the proper classes).

Definition 2.6. Let $s, t \in \mathcal{C}_{Ord}$ with $dom(s) = dom(t) = \alpha$. We define the slide of s into t as the function $\sigma_{st} : \mathcal{C}_{Ord} \longrightarrow \mathcal{C}_{Ord}$ as:

$$\sigma_{st}(x) = \begin{cases} x + (t - s), & \text{if } dom(x) = \alpha; \\ x + [(t - s) \upharpoonright \beta], & \text{if } dom(x) = \beta < \alpha; \\ x + [(t - s)\frown(\overline{0})_{<\beta-\alpha}], & \text{if } dom(x) = \beta > \alpha. \end{cases}$$

Theorem 2.2. For any $s, t \in \mathcal{C}_{Ord}$ with $dom(s) = dom(t) = \alpha$, the slide

$$\sigma_{st} : \mathcal{C}_{Ord} \longrightarrow \mathcal{C}_{Ord}$$

is an automorphism of $\langle \mathcal{C}_{Ord}, <_{Ord}, E \rangle$.

Proof. By straighforward verification, σ_{st} is 1-1 and onto.

To show that the order relation $<_{Ord}$ is preserved under σ_{st}, take $x <_{Ord} y$. This means that, for $\delta = min(dom(x), dom(y))$, $x \upharpoonright \delta <_\delta y \upharpoonright \delta$. Let $\alpha = dom(s) = dom(t)$. Whether $\alpha < \delta$ or $\alpha \geq \delta$, in any case, σ_{st} will modify by the same amount the individual components of x and y up to δ, so $\sigma_{st}(x) \upharpoonright \delta <_\delta \sigma_{st}(y) \upharpoonright \delta$ and therefore $\sigma_{st}(x) <_{Ord} \sigma_{st}(y)$.

As for the relation E, if xEy, then $x \upharpoonright \delta = y$, where $\delta = dom(y)$. As before, the action of σ_{st} up to δ, will equally act on x and y, so $\sigma_{st}(x) \upharpoonright \delta = \sigma_{st}(y)$, so we have that $\sigma_{st}(x)E\sigma_{st}(y)$. □

I conclude this section by emphasizing that the main properties of a continuum discussed in *Section 2.1* are realized by $\mathcal{C}_{Ord}$. According to the E relation (Definition 2.4) the model is non-well-founded, which means that it has no ultimate "elements" in the geometrical/mereological sense of "parts". There are no "points" given in advance that we could "put together" to form a set, so the model is *inextensible*. In agreement with Peirce's position, points are conceivable only as ideal limits of a process of growing specification.

This combines with Theorem 2.1 which addresses the requirement of *reflexivity*: each element $x \in \mathcal{C}_{Ord}$ has an inner structure (its associated monad $\mathcal{M}_x$) which reflects the totality of the model (through the obtained isomorphism). *Supermultitudinousness* holds obviously by the fact that $\mathcal{C}_{Ord}$ is a proper class, but we obtain also Peirce's requirement that between any two $<$-comparable elements there is any possible cardinal of elements (more than any "multitude of individuals", in Peirce's words).

The construction of $\mathcal{C}_{Ord}$ as a union of different stages $\mathcal{C}_\alpha$ may also be seen as a successive passage from the potential to the actual, "remembering that the word 'potential' means indeterminate yet capable of determination in any special case, there may be a potential aggregate of all possibilities that are consistent with certain general conditions".[49] This *modal* aspect of the model will be specified from a logical viewpoint in *Section 2.4*. I have already discussed how *genericity* is understood in the sense of homogeneity and as as the "absence of distinction of individuals".[50] Different aspects of this are to be seen in Theorem 2.1, Corollary 2.1 and Theorem 2.2.

The $\mathcal{C}_{Ord}$ construction also formalizes Peirce's proposal in *Subsection 2.1.3*. $\mathbb{R}$ is an initial stage reflecting that it is "the first embryo of continuity, ... an incipient cohesiveness, a germinal of continuity".[51] We then enrich this "line" by inserting over it, in an infinitary process, copies of it. This gives place to infinitesimals: already in $\mathcal{C}_2$ we obtain a non-Archimedean order but, in fact, it is possible to distinguish a whole hierarchy of different orders of infinitesimals (considering the construction associated to each ordinal). We may see the line as covered by "infinitesimal micro-segments" each of which, in its turn, may be covered in the same way (reflexivity). Infinitesimals are thus shown to be consistent, but may also play a role, in the specific version here obtained, in different mathematical fields. This will be explored later in *Sections 2.8* and *2.9*.

2.3 A Mereological View

> What is directly yielded to our knowledge by sense-awareness is a duration....
> A moment is a limit to which we approach as we confine attention to durations of minimum extension.
> *Whitehead 1920*

> The continuity of nature arises from extension.
> Every event extends over other events, and every event is extended over by other events...
> Thus there is no atomic structure of durations...
> *Whitehead 1920*

> Continuity concerns what is potential; whereas actuality is incurably atomic.
> *Whitehead 1929*

Parallels between Peirce and Whitehead have been noticed in different aspects of their thought.[52] Whitehead's process philosophy in fact involves a conception

---

49 [**1976**] NEM 3.106.
50 [**1931–58**] CP 4.172.
51 [**1976**] NEM 3.88.
52 See *e.g.* Nubiola 2008, or the wide-ranging study Brioschi 2020.

of the continuum with noticeable coincidences with Peirce's ideas. Whitehead's analysis of reality and natural phenomena are based on the idea that they are constituted primarily by processes and not by states. This leads to a conception of a geometrical space not constituted primarily by ultimate punctual entities, and therefore very close to what Peirce calls non-compositional, that is, non determinated by points.

Whitehead's ideas are exposed in an informal framework[53] but have led to several attempts of formalization, now constituted in the so called *point-free geometry*.[54] In this approach, the fundamental geometric concepts are redefined in terms of extended regions instead of indivisible points. With this in mind, I will contrast next the model $\mathcal{C}_{Ord}$ with some of the most fundamental axioms proposed for point-free geometry.

The relation E introduced before may be extended in a natural way to more general parts of the line, and not only to the elements of $\mathcal{C}_{Ord}$ (remember that, from Definition 2.4, xEy iff $dom(y) < dom(x)$ and $x \upharpoonright dom(y) = y$):

Definition 2.7. Given a and b, subsets of $\mathcal{C}_{Ord}$, let $\lambda_{a,b} = sup\{\mu | \mu = dom(s)$ for some $s \in a \cup b\}$. We will say that $a \sqsubseteq b$ iff for every $x \in \mathcal{C}_{(\lambda_{a,b}+1)}$, if xEa' for some $a' \in a$ then xEb' for some $b' \in b$. If both $a \sqsubseteq b$ and $b \sqsubseteq a$, we will say that a and b are equiextensive, and write $a \sim b$. In case that $a \sqsubseteq b$ but $a \nsim b$, we will write $a \sqsubset b$.

Let us also introduce an "overlapping" relation O as follows:

Definition 2.8. Given x and y, subsets of $\mathcal{C}_{Ord}$, we will say that xOy iff there is a $z \in \mathcal{C}_{Ord}$ such that $z \sqsubseteq x$ and $z \sqsubseteq y$.

The overlapping relation O satisfies the more characteristic property of mereological spaces according to Lesniewski:[55]

L2 $\forall z(zOx \to zOy) \to x \leq^M y$.

In our case the "pathood" or "inclusion" relation $\leq^M$ consists in the relation $\sqsubseteq$. The principle follows from the definition of the O relation: suppose that $x \nsqsubseteq y$; then there is a $z \in \mathcal{C}_{(\lambda_{x,y}+1)}$ such that zEx and $z\cancel{E}y$. Then zOx and $z\cancel{O}y$.

---

53 See Whitehead 1920, and Whitehead 1929.
54 See Gerla 2021 for a compilatory outlook on the topic.
55 Lesniewski 1916, as cited in Gerla 2021. Here I use the notation $\leq^M$ for the mereological relation, in order to avoid the multiple uses of ≤ already introduced before.

Gerla encapsulates the extensive list of principles provided by Whitehead in the following three axioms, where *Re* denotes a collection of "regions" (or "solids"):[56]

W1 $(Re, \leq^M)$ is a mereological space with no minimum and no maximum.
W2 $(Re, \leq^M)$ is dense in itself, *i.e.* if $x <^M y$ then there is z such that $x <^M z <^M y$.
W3 $\leq^M$ is upward-directed, *i.e.* for every x and y there is a z such that $x \leq^M z$ and $y \leq^M z$.

A model of these principles is called by Gerla a "Whitehead inclusion space" (WIS). Even so, we can speak of a WIS in a weak sense when the requirement for no maximum is removed. As Gerla himself states "it is secondary to assume that there is a maximum or not". Let us indicate with $\mathcal{P}(\mathcal{C}_{Ord})$ the class of *sets* of elements of $\mathcal{C}_{Ord}$. With this in mind we have that:

Theorem 2.3. $(\mathcal{P}(\mathcal{C}_{Ord}), \sqsubseteq)$ is a WIS with maximum.

Proof. As already shown, L2 follows directly from Definitions 2.7 and 2.8.

$\mathcal{P}(\mathcal{C}_{Ord})$ has no minimum because given an x in this class (*i.e.*, a subset of $\mathcal{C}_{Ord}$) it is always possible to find an element y of $\mathcal{C}_{\lambda_{x,x}+1}$ such that $y \sqsubset x$. This y may be obtained trivially by extending to a sequence of length $\alpha = \lambda_{x,x} + 1$ any sequence $x' \in x$, since the domain x', by definition of α, is $< \alpha$. It is clear that $\{y\} \sqsubset x$ and the inclusion is strict because we can construct a $z \in \mathcal{C}_{\alpha+1}$ such that $\{z\} \sqsubset x$ but $\{z\} \not E y$, simply extending the sequence x' arbitrarily with the restriction that, up to the the $\alpha - th$ component, this is done in a different way than previously done for y.

Now, for W2, let $x \sqsubset y$. This means that $x \not\to y$, so taking $\alpha = \lambda_{x,y} + 1$ there is a $w \in \mathcal{C}_\alpha$ such that wEy' for some $y' \in y$ but $w\not E x'$ for any $x' \in x$. We extend the α-sequence w with an arbitrary real and obtain a new sequence $w' \in \mathcal{C}_{\alpha+1}$. In order to define our z, we substitute w by w' in y, so $z = y \smallsetminus \{w\} \cup \{w'\}$. By definition of z we obtain that $x \sqsubset z \sqsubset y$, so we have the validity of W2.

For W3, given x and y subsets of $\mathcal{C}_{Ord}$, it suffices to take $z = x \cup y$, so trivially $x \sqsubseteq z$ and $y \sqsubseteq z$. □

By similar arguments, it is possible to obtain that different variants on the construction of $\mathcal{C}_{Ord}$, to be defined in the forthcoming sections, are also WIS, when an appropriate parthood relation $\sqsubseteq$ is introduced. For instance, $(\mathcal{C}_{Ord,Ord}, \sqsubseteq)$ (see *Section 2.5*) is a WIS (without maximum). Also the plane and in general n-dimentional continua (see *Section 2.6*) may be endowed with a WIS structure.

---

[56] Gerla 2021.

It is also important to notice how mereology as a general theory of parts regards the eventual existence of points. One of the problems posed in the context of point-free geometry, is precisely the definition of "point". The main approaches to this are inspired in Whitehead himself: "A point is the class of extended objects which, in ordinary language, contain that point".[57] These extended objects are "packed one within the other like the nest of boxes of a Chinese toy",[58] leading to a point as an ideal limit. We found already this conception in Peirce and have seen how the sequences in $\mathcal{C}_{Ord}$ lead to indefinitely precise approximations of ideal, purely potential points. In this sense nested loci defined by the E relation lead to "abstractive classes", even if a totality (a collection) of all these "points" is not an entity that can be defined.[59]

2.4 A Kripke Model View

The potential character of the continuum for Peirce was already emphasized in *Subsections 2.1.2* and *2.1.3*. The definition of the structure $\mathcal{C}_{Ord}$ passes through the stages of the $\mathcal{C}_\alpha$ in which the places are "indeterminate yet capable of determination in any special case". Also, we know that "given any collection of distinct individuals whatsoever, out of that potential aggregate there may be actualized a more multitudinous collection than the given one. Thus, the potential aggregate is, with the strictest exactitude, greater in multitude than any possible multitude of individuals".

In this section I will use the framework of Kripke models as a suitable context for formalizing these phenomena.[60] The use of Kripke models, of course a technical device introduced only some decades later, is nevertheless very close to Peirce's thought who talked about "possible worlds" and "universes of existence":

---

57 Cited in Varzi 2021, p. 353.
58 Whitehead 1920.
59 An appropriate definition of abstractive classes and points is not unproblematic, see Varzi 2021. In particular, in the context of $\mathcal{C}_{Ord}$, it deserves a further and more precise elaboration to be applied, specially regarding the need of proper classes in order to "reach" each specific "point".
60 As already proposed in Ketner and Putnam 1992. See Chellas 1980 or van Benthem 2010, for the basics on Kripke models.

What answers to our conception of a continuuum is a possibility of repeated division which can never be exhausted in any possible world, nor even in a possible world in which one can complete abnumerably infinite processes.[61]

The whole universe of true and real possibilities forms a continuum, upon which this Universe of Actual Existence is, by virtue of the essential Secondness of Existence, a discontinuous mark [...] There is room in the world of possibility for any multitude of such universes of Existence.[62]

The Kripke model is defined here as $\mathcal{M} = (Ord \smallsetminus \{0\}, R, \langle \mathcal{C}_{<\alpha}, \leq, E \rangle)$ where $Ord \smallsetminus \{0\}$ represents the possible worlds, the accessibility relation R is the order relation $\leq$ of the ordinals, and the models will be the $C_{<\alpha}$ considered, to start with, as first order structures.[63] We have some immediate consequences about the modalities in $\mathcal{M}$. For instance, since the relation R considered here is reflexive, we have that axiom T ($\Box A \to A$) holds, and since it is transitive, we have the validity of axiom 4 ($\Box A \to \Box \Box A$).[64]

The actual existence of elements of any cardinal may be expressed using cardinality quantifiers Q_κ.[65] An alternative approach for expressing the existence of elements at different cardinalities is to use arbitrarily long infinite chains of existential quantifiers and arbitrarily large infinitary conjunctions in the logic $\mathcal{L}_{\infty\infty}$. We have that $|\mathcal{C}_\alpha| = 2^{\aleph_0^{|\alpha|}}$ and $|\mathcal{C}_{<\alpha}| = sup\{2^{\aleph_0^{|\beta|}} : \beta < \alpha\}$. Therefore, for any cardinal κ, there is a least cardinal λ_κ,[66] such that

$$\mathcal{C}_{<\lambda_\kappa} \vDash Q_\kappa x(x = x)$$

and in general, for all $\lambda \geq \lambda_\kappa$,

$$\mathcal{C}_{<\lambda} \vDash Q_\kappa x(x = x).$$

---

[61] Ketner and Putnam 1992, p. 51.
[62] [1976] NEM 4.345.
[63] Other alternatives are possible: we could consider, for instance, the model on each world as C_α instead of $C_{<\alpha}$. I prefer here the choice of the last option because it allows to see each of these models as a partial approximation of the whole C_{Ord}.
[64] See Chellas 1980 p. 80. From similar considerations, it may be seen that Axioms 5 and B fail in $\mathcal{M}$.
[65] See Barwise and Feferman 1985.
[66] The exact value of λ_κ will depend on the behavior of the exponential function which (through the technique of forcing) may be subject to all the possibilities offered by Easton's Theorem, see Kunen 1980.

Since for any world there is another accessible world (remember that the accessibility relation is simply the usual order between ordinals) with at least κ elements, then for all λ and κ:

$$\mathcal{M} \Vdash_\lambda \Diamond Q_\kappa x(x = x).$$

Whereas for all $\lambda \geq \lambda_\kappa$,

$$\mathcal{M} \Vdash_\lambda \Box Q_\kappa x(x = x).$$

We therefore also have that for all λ and κ:

$$\mathcal{M} \Vdash_\lambda \Box\Diamond Q_\kappa x(x = x) \text{ and } \mathcal{M} \Vdash_\lambda \Diamond\Box Q_\kappa x(x = x).$$

This yields us some global truths in the whole model. In fact, for instance,

$$\mathcal{M} \vDash \Diamond Q_\kappa x(x = x)$$

but we also have stronger statements:

$$\mathcal{M} \vDash \Box\Diamond Q_\kappa x(x = x) \text{ and } \mathcal{M} \vDash \Diamond\Box Q_\kappa x(x = x).$$

We have that

$$\mathcal{M} \vDash Q_{2^{\aleph_0}} x(x = x)$$

and

$$\mathcal{M} \vDash \Box Q_{2^{\aleph_0}} x(x = x).$$

This reflects the *necessary* role assumed by the embryo of continuity, which, according to Peirce, is $\mathbb{R}$. In a way, the model $\mathcal{M}$ reflects the dynamics in the construction of $\mathcal{C}_{ord}$ which incarnates a world of possibilities beyond any cardinal, as stated in the quotes of *Subsection 2.1.2*.

We have also that in the model (as is usual with modal Kripke models) the principle of the excluded middle does not hold: for instance, it is neither true, nor false, that there are *actually* κ entities for $\kappa > 2^{\aleph_0}$:

$$\mathcal{M} \not\models Q_\kappa x(x = x)$$

and

$$\mathcal{M} \not\models \neg Q_\kappa x(x = x).$$

We have, though, also statements in plain first order logic for which the principle of the excluded middle fails in $\mathcal{M}$: $\forall x \exists y (yEx)$. In fact, the statement holds for limit ordinals, but fails for successor ordinals.[67]

We can also use $\mathcal{L}_{\infty\infty}$ for expressing the existence of arbitrarily long chains of nested loci:

$$\mathcal{C}_{<\alpha} \models \exists x_1 ... \exists x_\lambda ... \bigwedge_{\lambda+1<\alpha} x_{\lambda+1} E x_\lambda,$$

so, for every α,

$$\mathcal{M} \models \Diamond \exists x_1 ... \exists x_\lambda ... \bigwedge_{\lambda+1<\alpha} x_{\lambda+1} E x_\lambda.$$

As before, this leads to

$$\mathcal{M} \models \Box \Diamond \exists x_1 ... \exists x_\lambda ... \bigwedge_{\lambda+1<\alpha} x_{\lambda+1} E x_\lambda$$

and

$$\mathcal{M} \models \Diamond \Box \exists x_1 ... \exists x_\lambda ... \bigwedge_{\lambda+1<\alpha} x_{\lambda+1} E x_\lambda.$$

Summing up, we have seen different examples of truths that hold from a world onward in the Kripke model, giving rise to necessary statements as well as to necessary possibilities and possible necessities. This may be seen as a manifestation

---

[67] As discussed at the end of *Subsection 2.1.1*, Peirce conceived the nature of the continuum as challenging well established traditions in reasoning such as the principle of the excluded middle. In fact, in the original continuum "[...] the principle of excluded middle, or that of contradiction, ought to be regarded as violated" [**1976**] 3.747.

of the Peircean concept of knowledge as convergence of what stabilizes "in the long run".

2.5 From the Infinitely Small to the Infinitely Large

In this section, I will introduce a variant[68] of $\mathcal{C}_{Ord}$ which allows us to extend the construction, not only in the direction of the "infinitely small" but also in the opposite direction. It arises from reflecting on the way the usual decimal notation works.[69] Let us come back to the meaning of the construction performed in the preceding section. $\mathcal{C}_{Ord}$ is defined taking series of real numbers with different ordinal length. In these series, each real number which appears is a piece of information which specifies with increasing precision a place, a position in a line. In other terms, the E relation establishes that a series s specifies a place, and that a longer series extending it $t = s\widehat{\;}s'$ specifies a place "inside" s (so tEs).

This is entirely analogous with the positional representation system of numbers. Let us take, for instance, as our straight line, the closed interval [0, 10] and let's suppose we want to specify a point[70] on it: if we have only the information of the digit representing the units, then we can only specify 10 positions and we can think this as a line (let's call it r_1) with only such positions. Let's suppose for instance that the value of this digit is 6. We can acquire new information and know that the next digit is, say, 3. We place this new digit at the right and obtain 6, 3. This gives more precision to the position where the point is placed. We can interpret this as a new line r_2 formed by the series of 2 digits. The space is formed here by 100 positions. We can repeat this process as far as we want to specify the place. So, 6, 3276 is a place in a line r_5 which is more fine-grained than the previous ones.

As we have seen in *Section 2.2*, this is something similar to what happens in the construction of $\mathcal{C}_{Ord}$. Instead of adding to our sequences new digits as we have just done, we form the sequences by adding new real numbers. Also, instead of

---

68 Defined in Vargas 2015.
69 It is worth noticing that Peirce himself thought about possible extensions of the decimal notation. For instance, he considers the use of an ω-th position: "...two values, that differ at all, differ by a finite value, which would not be true if the ω-th place of decimals were supposed to be included in their exact expressions; and indeed the whole purpose of the doctrine of limits is to avoid acknowledging that that place is concerned" [**1931–58**] CP 6.176.
70 Here I use the word in its usual sense.

having to the right ω positions to occupy with new digits, we have, in $\mathcal{C}_{Ord}$, α positions to occupy with real-numbers, for every ordinal α (so, we could say that we have "*Ord*" possible positions to occupy). We are also changing the "line" or the model in which we are located, obtaining more and more specific possible "places". We spoke earlier, in this sense, of different pseudo-continuums of order α.[71]

This also reflects the idea that we can have different types of infinitesimals and that being infinitesimal is not something intrinsic to a number or a geometric entity, but is something relative, "with respect to". As Leibniz posed it, we can have "different degrees of infinity and of infinitelly small".[72] Thus, in $\mathcal{C}_{Ord}$, the 3-sequence $(0; 0; 2)$ is infinitesimal with respect to the 2-sequence $(0; \pi)$, which in turn is infinitesimal with respect to the 1-sequence (47).

Nothing prevents us to think as well about the possibility of different degrees of infinity ("infini de l'infini", again according to Leibniz) with some natural modifications on the construction of $\mathcal{C}_{Ord}$. This is possible if we follow as before the analogy with the positional system: we can in fact also add digits to the left of our 6 above. If we add the digit 2, we obtain a 26 that tells us that actually our initial line r_1 is part of a longer line, let's call it $r_{1;1}$, which is a union of several copies of r_1, in this case 10 of them. Now, the information conveyed by the 2 specifies not "what is inside" a given place but "within what" is located that given place.

The role played by "left" and "right" in the decimal representation of the reals is not completely symmetric, however. The decimal part (to the right of the comma) can indeed be of infinite length, while the integer part (to the left) is always finite. This finite character allows us to compare real numbers and see, given two of them, which one is greater than the other one. We can also modify the construction of $\mathcal{C}_{Ord}$ extending it "to the left", giving us different degrees of "infinity". We can do this in two ways: either by considering the sequences as starting on the right and developing to the left, or, conversely, starting at the left and developing to the right. In the case of finite sequences the difference is only apparent but in the general case of α-sequences for any ordinal α the difference is substantial, since in the last case (development from right to left), we cannot define anymore the lexicographic order which was crucial in $\mathcal{C}_{Ord}$. In this case, we can have different sequences from which it is not possible to determine which one is greater than the other one. We also lose the possibility of thinking places as being nested

---

[71] The decimal representation here is analogous to the one used for Hyperreal numbers, in the sense that it generalizes the real decimal positions beyond the natural numbers. Both representations differ, nevertheless, by the fact that the new positions that appear in Hyperreals correspond to the "Hypernaturals", whereas here the extension ranges along the ordinal numbers.
[72] "...plusieurs degrés d'infini ou infiniment petits..."

within other places as expressed by the E relation. Consider for instance the alternating ω-sequences s and t defined by $s_n = (-1)^n$ and $t_n = (-1)^{n+1}$. If we would take them increasing "from right to left" then we cannot compare them.

In what follows, we will consider the construction of $\mathcal{C}_{\vec{\alpha};\beta}$ in which the sequences increase "from left to right". A separate development, as can be seen from the previous considerations, deserves the case in which they go from right to left ($\mathcal{C}_{\overleftarrow{\alpha};\beta}$) which I will not treat here. Having said this, since there is no danger of ambiguity, I will omit the arrows above α from now on. As the reader will have noticed, another aspect not explored here would be to study the two directions of growth also with respect to the infinitesimal part. We then have the four possible combinations $\mathcal{C}_{\vec{\alpha};\vec{\beta}}$, $\mathcal{C}_{\vec{\alpha};\overleftarrow{\beta}}$, $\mathcal{C}_{\overleftarrow{\alpha};\vec{\beta}}$ and $\mathcal{C}_{\overleftarrow{\alpha};\overleftarrow{\beta}}$, of which I will consider only the first one.

Definition 2.9. For any pair of ordinals $\alpha, \beta \in Ord \smallsetminus \{0\}$, $\langle \mathcal{C}_{\alpha;\beta}, <_{\alpha;\beta}\rangle$ is defined as the set of all the ordered pairs $(s; s')$ where s is an α-sequence and s' a β-sequence of real numbers. The order $<_{\alpha;\beta}$ is lexicographic:

$$(s; s') <_{\alpha;\beta} (t; t') \Leftrightarrow (s <_\alpha t) \vee (s = t \wedge s' <_\beta t')$$

where $<_\alpha$ and $<_\beta$ are the corresponding order relations of $\mathcal{C}_\alpha$ and $\mathcal{C}_\beta$.

Continuing with the idea of writing the elements corresponding to a greater degree of infinity "positionally" further to the left, we can write the elements $(s; s')$ of $\mathcal{C}_{\alpha;\beta}$ as a single sequence like this:

$$s_0; s_1; \ldots; \underline{s'_0}; s'_1; s'_2; \ldots$$

In this notation, the element s'_0 corresponding to the finite part of our sequence is underlined. The elements of s, to the left of s'_0, indicate the real numbers at the different infinitary positions. To the right of s'_0 we have the remainder of the sequence s' that extends through the different degrees of infinitesimals covering the whole length β.

As an example, let us consider the finite sequence $s = (23; 44; \sqrt{5})$, and let a s' be defined as follows:

$$s'_\delta = \begin{cases} -276, & \text{for } \delta = 0; \\ \cos(7777), & \text{for } \delta = 1; \\ \pi, & \text{for } 1 < \delta < \beta = \omega_2; \end{cases}$$

We can indicate $(s; s') \in \mathcal{C}_{3;\omega_2}$ as :

$$23; 44; \sqrt{5}; \underbrace{-276}; \cos(7777); \overbrace{\pi; \pi; \ldots}^{\omega_2 \text{ times}}$$

The definition of the E relation extends to $\mathcal{C}_{\alpha;\beta}$, as well as the results of the previous sections which are transferred naturally. Basically, what we do is a recoding of the $\mathcal{C}_\alpha$ structures interpreting them in another way. An isomorphism can be established straightforwardly:

Theorem 2.4.
$$\langle \mathcal{C}_{\alpha;\beta}, <, E \rangle \simeq \langle \mathcal{C}_{\alpha+\beta}, <, E \rangle.$$

Finally, we can also consider the proper class that encompasses all of this type of sequences:

Definition 2.10.
$$\mathcal{C}_{Ord;Ord} = \bigcup_{\alpha,\beta \in Ord \smallsetminus \{0\}} \mathcal{C}_{\alpha;\beta}.$$

Other constructions, such as the indicated by $\mathcal{C}_{<\alpha;<\beta}$ and other variations have natural straightforward definitions.

2.6 From the Line to the Plane and Beyond

In the previous sections I have limited myself to the construction of a 1-dimensional continuum. Nevertheless, it is possible to think in higher dimensions, as many of the applications and intuitions about continuity demand. There are three possible natural approaches to define a bidimensional (or in general λ-dimensional for any ordinal λ) continuum based on what has been done with $\mathcal{C}_{Ord}$. In order to fix the ideas, let us consider the bidimensional case:

- It is possible to repeat the construction done with $\mathcal{C}_{Ord}$, but instead of using elements of $\mathbb{R}$ in order to construct sequences, we can use elements of $\mathbb{R}^2$ (see *Figure 2.2*). What emerges then is entirely analogous to the 1-dimensional case: for the case of dimension 2, for instance, what we have is an initial first approximation to the usual plane $\mathbb{R}^2$. In the next step what we see is that each apparent point of $\mathbb{R}^2$ has inside it an entire copy of $\mathbb{R}^2$, etc. (producing sequences of couples of length α: $^\alpha(\mathbb{R}^2)$).
- We can also start by the construction of $\mathcal{C}_\alpha$ as before, and construct the set of pairs of elements on it: $(\mathcal{C}_\alpha)^2 = \mathcal{C}_\alpha \times \mathcal{C}_\alpha$ (see *Figure 2.3*). Then we can take the union of these.

- A third way would be to take all possible ordered pairs of $\mathcal{C}_{Ord}$, namely $(\mathcal{C}_{Ord})^2 = \mathcal{C}_{Ord} \times \mathcal{C}_{Ord}$.

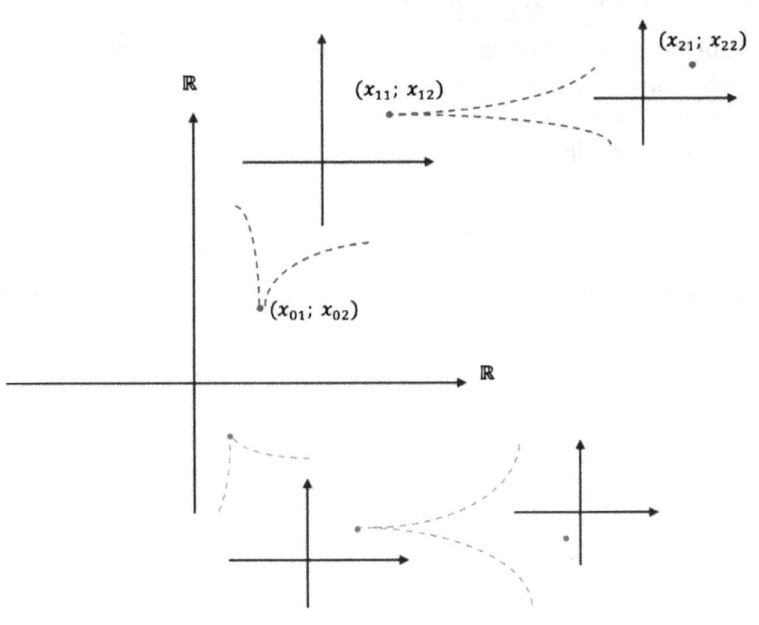

Figure 2.2: $^3(\mathbb{R}^2)$ is structured, giving rise to places such as $((x_{01}; x_{02}); (x_{11}; x_{12}); (x_{21}; x_{22}))$.

Theorem 2.5. The following constructions are isomorphic:
1. $\langle \bigcup_{\alpha \in Ord}{}^{\alpha}(\mathbb{R}^2), E \rangle$
2. $\langle \bigcup_{\alpha \in Ord}(\mathcal{C}_\alpha)^2, E \rangle$

Proof. The isomorphism follows considering an arbitrary element $x \in {}^{\alpha}(\mathbb{R}^2)$. We have that x is of the form $((x_{01}; x_{02}); (x_{11}; x_{12}); \ldots; (x_{\lambda 1}; x_{\lambda 2}); \ldots)$ with $\lambda < \alpha$. We can establish an assignment: $f: {}^{\alpha}(\mathbb{R}^2) \longrightarrow (\mathcal{C}_\alpha)^2$ defined by

$$x \longmapsto ((x_{01}; x_{11}; \ldots; x_{\lambda 1}; \ldots); (x_{02}; x_{12}; \ldots; x_{\lambda 2}; \ldots)).$$

It is easily verified that the E relation is preserved. □

The equivalence allows us to conceive the plane interchangeably from two different intuitions. We can consider the construction $\mathcal{C}_{Ord}$ or partial approximations to it ($\mathcal{C}_{<\alpha}$) on both axis, so we can consider functions (or relations) defined on dif-

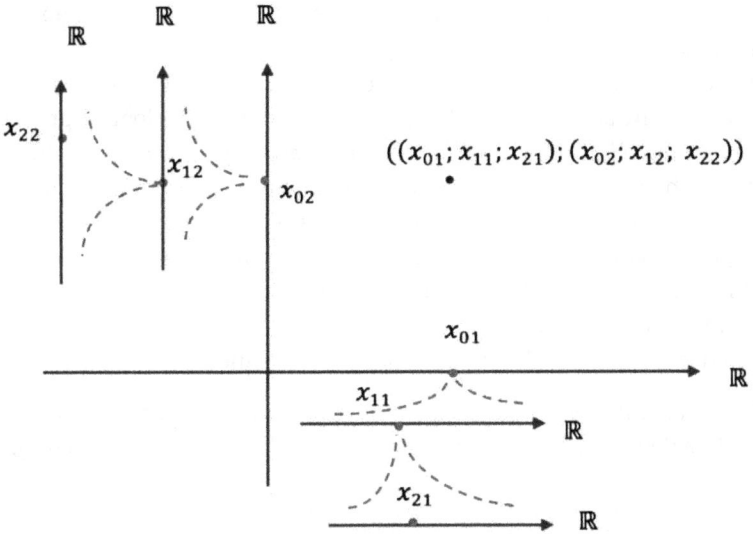

Figure 2.3: In the case of $\mathcal{C}_3^2$, we start first by considering sequences of $\mathcal{C}_3$, forming then ordered pairs such as $((x_{01}; x_{11}; x_{21}); (x_{02}; x_{12}; x_{22}))$.

ferent strata of these constructions. At the same time we can view each "point" of the plane as having inside a copy of the whole plane.

We notice that the third construction above is strictly larger, since $\bigcup_{\alpha \in Ord}(\mathcal{C}_\alpha)^2$ contains pairs of sequences of the same length, whereas in $(\mathcal{C}_{Ord})^2$ we obtain also pairs of sequences of different domains.

2.7 An Algebraic View

Peirce, like other authors, repeatedly emphasized the purely qualitative character of the "primordial continuum" (to return to Thom's expression.[73]) According to Peirce, "Number cannot possibly express continuity",[74] and "There will be lengths not measurable by [...] numbers, nor by limits of series of them".[75] Of course, he had in mind the number systems known at his time and specially

---

[73] See footnote 33.
[74] [1976] NEM 3.93.
[75] [1976] NEM 3.127.

the real-numbers approach to define "the line", as well as the identification of numbers with points.

In the previous sections the definitions of C_{Ord} and its variants have been formulated in terms of the study of its order properties and its mereological properties. However, the model also has a whole algebraic structure inherited naturally from $\mathbb{R}$. This can be seen as an approach to the continuum through a generalized notion of number encompassing different degrees of infinitesimals and, in the case of $C_{Ord;Ord}$, going also into the direction of the infinitely large. Following Vargas,[76] in this section I will focus on the algebraic structure of C_{Ord}. Natural adaptations can be done also for $C_{Ord;Ord}$.

As I showed in *Section 2.2*, the sum in C_{Ord} can be defined componentwise:

Definition 2.11. Consider two sequences $s, t \in C_{Ord}$ with $\alpha = dom(s)$ and $\beta = dom(t)$. We define their sum $s+t \in C_\mu$ componentwise for each $y < \mu = max(\alpha, \beta)$:

$$(s+t)_y = \begin{cases} s_y + t_y, & \text{for } y < min(\alpha, \beta); \\ s_y, & \text{if } \beta < y < \alpha; \\ t_y, & \text{if } \alpha < y < \beta. \end{cases}$$

Now, let us turn to the definition of a product. Let us remind first the following result by Cantor[77] that gives us a canonical way of representing ordinals:

Theorem 2.6. (Cantor Normal Form-CNF) Every ordinal number α can be written in a unique way in the form

$$\omega^{\beta_1} a_1 + \omega^{\beta_2} a_2 + \ldots + \omega^{\beta_k} a_k$$

where k and $a_1, \ldots, a_k$ are natural numbers and $\beta_1 > \beta_2 > \ldots > \beta_k$ are ordinal numbers.

Making use of this representation we can define a product in the model, which generalizes the multiplication of polynomials in one variable:

Definition 2.12. Given two ordinal numbers α and β let us represent them in their CNF:

$$\alpha = \omega^{\alpha_1} a_1 + \omega^{\alpha_2} a_2 + \ldots + \omega^{\alpha_k} a_k$$

---

76 Vargas 2015.
77 See *e.g.* Sierpiński 1958.

$$\beta = \omega^{\beta_1} b_1 + \omega^{\beta_2} b_2 + \ldots + \omega^{\beta_l} b_l.$$

We define the ordinal:

$$\alpha \oplus \beta = \omega^{\gamma_1} c_1 + \omega^{\gamma_2} c_2 + \ldots + \omega^{\gamma_k} c_m$$

where the ordinal numbers $\gamma_1, \gamma_2, \ldots, \gamma_m$ are all the α_i and β_j given by the representations of α and β, taken in increasing order and without repetitions. The natural numbers c_n are defined by:

$$c_n = \begin{cases} a_i + b_j, & \text{if } \gamma_n = \alpha_i = \beta_j; \\ a_i & \text{if } \gamma_n = \alpha_i \text{ but doesn't appear in the CNF of } \beta; \\ b_j & \text{if } \gamma_n = \beta_j \text{ but doesn't appear in the CNF of } \alpha. \end{cases}$$

Definition 2.13. Given $s, t \in \mathcal{C}_{Ord}$ with $\alpha = dom(s)$ and $\beta = dom(t)$, we define the product $p = s \odot t \in \mathcal{C}_{\alpha \oplus \beta}$ as the sequence defined for each $\gamma < \alpha \oplus \beta$ by:

$$p_\gamma = \sum_{\substack{\sigma \oplus \tau = \gamma \\ \sigma < \alpha \\ \tau < \beta}} s_\sigma \cdot t_\tau.$$

Notice that the product $p = s \odot t \in \mathcal{C}_{Ord}$ because in fact each p_γ is a real number. This holds thanks to the fact that the sum in the definition has finite character. In fact, each γ can be written (in a unique way) in the CNF, say, as $\omega^{\gamma_1} c_1 + \omega^{\gamma_2} c_2 + \ldots + \omega^{\gamma_k} c_m$, and given that the c_n are natural numbers, there is only a finite number of possible ordinal values σ and τ such that $\sigma \oplus \tau = \gamma$, by the definition of $\oplus$.

In the case of the 1-sequences, we have just a real number and the product $\odot$ is equal to the usual product of real numbers. In fact, in this case, in the sum of Definition 2.13, $\alpha = \beta = 1$ and the only possible value of σ and τ is 0. Therefore $\sigma \oplus \tau = 0$ and the only p_γ is p_0, which means that we obtain as a result a real number. In other words, taking $p = s \odot t$, then p is a 1-sequence whose only element is $p_0 = s_0 \cdot t_0$.

Another desirable property of this product is that, with the appropriate definition of powers, when raising infinitesimals to different powers we obtain different degrees of infinitesimals. In fact, given an infinitesimal ϵ, we have $\epsilon \gg \epsilon^2 \gg \epsilon^3 \ldots$ where the relation $a \ll b$ indicates that a is infinitesimal for b. In the event that we do not work in all $\mathcal{C}_{Ord}$, but limit ourselves to certain C_α, by this property we can obtain *nilpotent* elements, as in other contexts like Smooth Infinitesimal Analysis.[78]

It can be routinely checked that the $\odot$ operation is associative and commutative. We prove its distributivity with respect to the sum:

Theorem 2.7. For all $s, t, t' \in \mathcal{C}_{Ord}$ we have that

$$s \odot (t + t') = s \odot t + s \odot t'.$$

Proof. Consider some $s, t, t' \in \mathcal{C}_{Ord}$ with $dom(s) = \alpha$, $dom(t) = \beta$ and $dom(t') = \beta'$. Without losing generality suppose that $\beta \geq \beta'$. We define the sequences

$a = s \odot (t + t')$, and
$b = s \odot t + s \odot t'$.

Let us see that $a_y = b_y$ for all $y < \alpha \oplus \beta$.

We have that

$$a_y = \sum_{\substack{\sigma \oplus \tau = y \\ \sigma < \alpha \\ \tau < \beta}} s_\sigma \cdot (t + t')_\tau.$$

Given that $y < \beta'$ then by the definition of the sum between sequences this sum may be expressed as:

$$\sum_{\substack{\sigma \oplus \tau = y \\ \sigma < \alpha \\ \tau < \beta}} s_\sigma \cdot (t_\tau + t'_\tau).$$

In those cases in which t'_τ is not defined, the expression $t_\tau + t'_\tau$ reduces only to t_τ.

By distributivity between real numbers and properties of sums we obtain finally:

$$\sum_{\substack{\sigma \oplus \tau = y \\ \sigma < \alpha \\ \tau < \beta}} s_\sigma \cdot t_\tau + s_\sigma \cdot t'_\tau = \sum_{\substack{\sigma \oplus \tau = y \\ \sigma < \alpha \\ \tau < \beta}} s_\sigma \cdot t_\tau + \sum_{\substack{\sigma \oplus \tau = y \\ \sigma < \alpha \\ \tau < \beta}} s_\sigma \cdot t'_\tau = \sum_{\substack{\sigma \oplus \tau = y \\ \sigma < \alpha \\ \tau < \beta}} s_\sigma \cdot t_\tau + \sum_{\substack{\sigma \oplus \tau = y \\ \sigma < \alpha \\ \tau < \beta'}} s_\sigma \cdot t'_\tau$$

This last expression is precisely $(s \odot t + s \odot t')_y = b_y$. $\square$

---

78 Bell 1998. See below *Sections 2.8* and *2.9*.

We can define in $\mathcal{C}_{Ord}$ the equivalence relation given by:

$s \sim t \Leftrightarrow s$ and t have the same support and have the same values when restricted to it.

This means that whether the sequences s and t have the same length or not, in order for them to be equivalent they must equate in their non-zero elements. From the previous results we conclude that

Corollary 2.2. $\mathcal{C}_{Ord}/\sim$ has a ring structure, with the operations $+$ and $\odot$ defined between the equivalence classes previously defined.

The study of the algebraic structure of $\mathcal{C}_{Ord}$ and its variants may have different refinements. In particular, something unexplored yet are the relations with other infinitesimalist constructions such as Levi-Civita's,[79] Hahn's,[80] and Ehrlich's.[81] See Ehrlich,[82] for a wide overview on the topic.

2.8 A Geometric View

It is possible to use the definitions so far developed in order to introduce geometrical notions through an algebraic geometry approach. Curves in the plane (or manifolds in α-dimensional spaces) are defined as solutions of equations. Coefficients and solutions may be considered as belonging to the different levels of $\mathcal{C}_{Ord}$. This natural path will not be developed here.

I will limit myself to remark that the algebraic definitions in the previous section generalize the structure of the dual numbers, used by Johannes Hjelmslev as a way to provide a model for his "geometry of reality".[83] Dual numbers are denoted by $\mathbb{R}[\varepsilon]$, and are constituted by numbers of the kind $a+\varepsilon b$ with $a, b \in \mathbb{R}$, where ε is an infinitesimal (therefore $\neq 0$) such that $\varepsilon^2 = 0$. They may be presented then as ordered pairs, or 2-sequences of the form $(a; b)$, and thus, as members of $\mathcal{C}_2$. We must add the additional condition that if $a = 0$ then we obtain nilpotent elements. In other words, in the context of the model $\mathcal{C}_{Ord}$, dual numbers may be identified with the structure $\langle \mathcal{C}_2, \leq, +, \odot \rangle$ where $\odot$ is redefined as an operation $\odot : \mathbb{R}^2 \longrightarrow \mathbb{R}^2$

---

[79] Levi-Civita 1893.
[80] Hahn 1907.
[81] Ehrlich 2010.
[82] Ehrlich 2021.
[83] Lützen 2021.

such that infinitesimal elements are nilpotent. In this way we obtain (from the previous section) a commutative ring with identity.[84]

Geometry developed from dual numbers allows for two different incident lines to have in common more than a single point. So the Euclidean axiom for the uniqueness of a line passing through two different points fails. In fact, failure of this principle characterizes what are known as Hjelmslev (incidence) geometries.[85] We will see in the next section that it is also possible to develop Analysis through an infinitesimal approach using dual numbers. Even so, I will not restrict myself to their use, leaving the space open to different extensions and generalizations of the concepts to the different levels of $\mathcal{C}_{Ord}$ (beyond $\mathcal{C}_2$).

Another possible path of research in this context, more connected to Peirce's characteristic themes, is the topological approach to logic via his Existential Graphs. These were considered by him a foremost application of continuity, so it should be natural to see which consequences or possibilities may be opened if their underlying plane is not the usual one ($\mathbb{R}^2$), but a plane reflecting his own ideas on the continuum. Here we may follow Oostra's works extending the use of EGs beyond what Peirce himself proposed[86] (in some cases because the logics required did not exist at the time). With the use of a $\mathcal{C}_{Ord}$-based plane, we may think in an enrichment of expressive power which allows infinitary conjunctions and infinitary chains of quantifiers for any cardinal (given the supermultitudinousness of the model). Also, the layered structure of the places on the plane can be interpreted as going beyond first-order logic, to higher order ones. Quantification over these different orders is rendered possible by identity lines with different "widths". These may be obtained by lines defined on different degrees of resolution of the plane.

2.9 Infinitesimal Calculus: an Alternative Approach

Using the ideas from the previous sections, we can obtain traditional Calculus results through the use of infinitesimals, as an alternative to other infinitesimal approaches such as Non-Standard Analysis (NSA) and Smooth Infinitesimal Analysis (SIA). Many of the definitions and results of this section are closer to the spirit

---

84 See Ehrlich 2021 and Lützen 2021.
85 Veldkamp 1995.
86 See Oostra's *Chapter 3* in this book.

of SIA as developed, *e.g.*, in Bell.[87] In fact we can construct definitions (at least in the case of elementary functions) in such a way that the "Principle of Micro-Straightness" works.

Using the equivalence in Theorem 2.5, we can use different intuitions about the plane. On one hand, we use the definition of the plane as

$$\mathcal{P} = \bigcup_{\alpha \in Ord \smallsetminus \{0\}} (\mathbb{R}^2)^\alpha.$$

Let us remind that given sequences $x, y \in \mathcal{P}$, we will say (in a similar way as in Definition 2.4), that xEy iff $dom(y) < dom(x)$ and $x \upharpoonright dom(y) = y$. On the other hand, we can consider functions defined on $\mathcal{C}_{<\alpha}$ as being constituted by a set of α strata of functions defined on the levels $\mathcal{C}_\beta$ for $\beta < \alpha$. In reality, we will need for the sake of the forthcoming presentation only two levels of resolution since most of the arguments can be recodified in terms of $\mathbb{R}^{\leq 2}$. So $\mathcal{C}_{\leq 2}$ will be enough for performing infinitesimal arguments, at least in a first approximation.

It is generally assumed[88] that there are two main approaches to deal with infinitesimals, namely, considering them as either invertible or nilpotent. In our case of $\mathcal{C}_{<\alpha}$ or $\mathcal{C}_{Ord}$, we have that infinitesimals are not invertible, nor a priori nilpotent. We can obtain invertible infinitesimals using the $\mathcal{C}_{\alpha,\alpha}$ and $\mathcal{C}_{Ord,Ord}$ constructions of Definitions 2.9 and 2.10. In this case, of course, inverses of infinitesimals will be constituted by infinite elements. This allows to develop analysis tools similar to the used in NSA, or even closer, to the approach in Tall.[89] The "superreal numbers", which admit invertible infinitesimals, are defined there as formal power series and may be in fact seen as isomorphic to $\mathcal{C}_{<\omega,\leq\omega}$.

In this section we follow another approach: we will have the macroscopic level of the plane $\mathbb{R}^2$, where functions are defined in the standard way, and the infinitesimal level is specified in $(\mathbb{R}^2)^2$. Given an infinitesimal $\varepsilon \in \mathcal{C}_2$, we would have by the definition of the product, that $\epsilon^2 \in \mathcal{C}_3$. Since $\varepsilon^2 \ll \varepsilon$, we may ignore this value, so we can define now that $\varepsilon^2 \approx 0$. If instead of approximation we take equality, this turns out to be very close, in practice, to the use of nilpotent elements which are used in SIA (and in Hjelmslev Geometries, mentioned in the previous section).

Also, as in SIA, we can obtain *micro-straightness*. In the present context, micro-straightness is not obtained in general: functions can be defined arbitrarily without this property, as we will see. Nevertheless, given a real-valued function f,

---

[87] Bell 1998.
[88] Ehrlich 2021.
[89] Tall 1979 and Tall 1980.

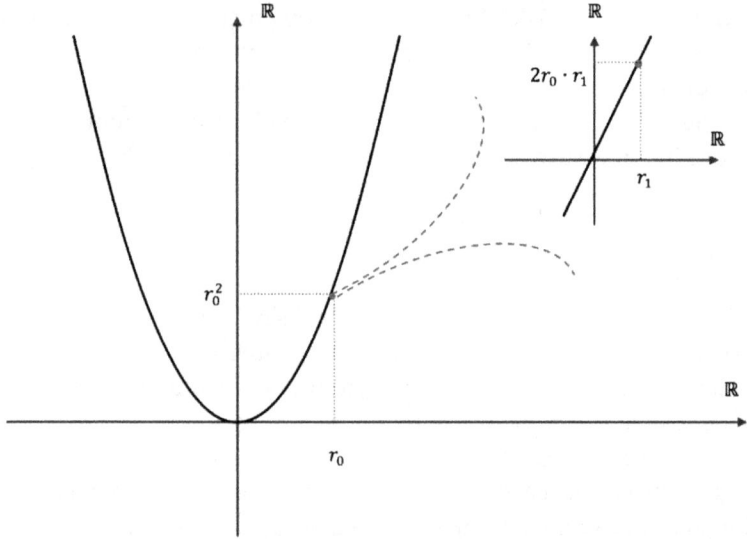

Figure 2.4: Microstraightness behavior: as an example, consider here the function $f(x) = x^2$ as a $\mathcal{C}_2$-function defined by the expression $f(r_0; r_1) = (r_0; r_1) \odot (r_0; r_1) = (r_0^2; 2r_0 \cdot r_1)$. Inside the place $(r_0; r_0^2)$, r_0 is fixed and the expression $2r_0 \odot r_1$ is linear in r_1, precisely with coefficient $2r_0$, the derivative of the function. See *Subsections 2.9.3* and *2.9.4* for formal definitions.

it is possible always to assign a function $^*f : \mathcal{C}_2 \to \mathcal{C}_2$ which is micro-straight. In the case of the elementary functions explored next, micro-straightness will follow from their defining formula. This will allow us to introduce derivatives and integrals in a natural way in these cases. Micro-straightness implies differentiability and local continuity. Nevertheless a much wider class of functions is conceivable which are continuous in the NSA sense,[90] but that nevertheless should not be called continuous according to the Leibnizean intuition of absence of jumps.[91] For instance, intermediate value theorems in general do not hold.

Let us consider next a general framework on properties that, in the present context, the behavior of functions may possess or not. These turn out to hold in the case of elementary functions where micro-straightness is obtained.

---

[90] See Cauchy's criterion in Goldblatt 1998, p. 76, and Section 2.9.1.
[91] *Natura non facit saltus.*

2.9.1 Functions

In order to define functions in the context of $\mathcal{C}_{Ord}$, different approaches may be considered. In fact, in the general case functions are arbitrary and we can assign images with no geometric constraint. Nevertheless, we can impose coherence conditions between the different "degrees of resolution" that we obtain along the different levels of $\mathcal{C}_\alpha$. This is natural in particular in the case in which we start with a real-valued function. Thus we have a macro-level which can have different kinds of developments at the micro-levels $\mathcal{C}_\alpha$ for ($\alpha > 1$).

Let us consider for instance a constant function $f : \mathbb{R} \to \mathbb{R}$ given by $f(x) = k$. We may think in extending the function through its values at $\mathcal{C}_\alpha$ for ($\alpha > 1$) using the "analytical expression" already defined for $\mathbb{R}$. Several examples of this process will be seen in *Subsection 2.9.3*. But many alternatives to this analytic extension are also possible. We may have that f is constant at the macro-level of $\mathbb{R}$ but non-constant at the finer grained level of, say, $\mathcal{C}_2$. We may even have that at this level the function is not continuous, as the next example shows.

Example 2.1. Let us define an extension of the function $f(x) = k$ just given. Let us indicate it as $^*f : \mathcal{C}_2 \to \mathcal{C}_2$ an let it be given (using Dirichlet-style functions) by:

$$^*f(r_0; r_1) = \begin{cases} (k; 1) & \text{if } r_1 \in \mathbb{Q} \\ (k; 0) & \text{if } r_1 \notin \mathbb{Q}. \end{cases}$$

This kind of examples will force us to define different levels of continuity. It will not be enough in general to use Cauchy's criterion of continuity as used in NSA.[92] According to Cauchy:

> the function $f(x)$ is continuous with respect to x between the given limits if between these limits an infinitely small increase in the variable always produces an infinitely small increase in the function.

This criterion is satisfied by the function just defined, but, as I mentioned above, a different definition is needed to caracterize the intuition of continuity at different levels.

We can also have the opposite case: a function which is discontinuous at the macro-level of the reals, but continuous at the resolution of $\mathcal{C}_2$:

---

[92] See Goldblatt 1998, p. 75.

Example 2.2. Take $g : \mathbb{R} \to \mathbb{R}$ as the usual Dirichlet function, and extend it by:

$$^{*_1}g(r_0; r_1) = \begin{cases} (1;0) & \text{if } r_0 \in \mathbb{Q} \\ (0;0) & \text{if } r_0 \notin \mathbb{Q}. \end{cases}$$

In this case g is discontinuous at every real number, but $^{*_1}g$ appears to be intuitively continuous (and differentiable) at every point of its domain.

Example 2.3. We can even have a function which is intuitively discontinuous at both levels but which is continuous according to Cauchy's criterion, namely, starting with the g above, and defining this time:

$$^{*_2}g(r_0; r_1) = \begin{cases} (1;1) & \text{if } r_0 \in \mathbb{Q} \text{ and } r_1 \in \mathbb{Q} \\ (1;0) & \text{if } r_0 \in \mathbb{Q} \text{ and } r_1 \notin \mathbb{Q} \\ (0;1) & \text{if } r_0 \notin \mathbb{Q} \text{ and } r_1 \in \mathbb{Q} \\ (0;0) & \text{if } r_0 \notin \mathbb{Q} \text{ and } r_1 \notin \mathbb{Q}. \end{cases}$$

The previous examples can illustrate, even at the initial stage of $\mathcal{C}_2$, the extreme diversity of possibilities on how real-valued functions can be extended to the infinitesimal levels. Nevertheless, in the following, I will limit myself to consider the case of some of the usual elementary functions and how the concepts of calculus may be defined on them.

2.9.2 Coherence, Stratification, and Continuity

Cauchy's definition of continuity mentioned in the last subsection can be generalized and refined considering different degrees of infinitesimals. This way, infinitesimals at the different micro-levels in the domain of a function do not affect what occurs in higher levels but only at their own levels (or lower). This can be formalized in general, extending Cauchy's principle:

Definition 2.14. $f : \mathcal{C}_{<\beta} \to \mathcal{C}_{Ord}$ is coherent iff for every $s \in \mathcal{C}_{<\beta}$, if $t \in \mathcal{M}_s \cap \mathcal{C}_{<\beta}$ then $f(t) \in \mathcal{M}_{f(s)}$. In other words, if tEs then $f(t)Ef(s)$.

Generally speaking, functions defined by an analytic expression are coherent. Nevertheless, counterexamples abound considering more arbitrary functions:

Example 2.4. Consider for instance the function $f : \mathcal{C}_{<3} \to \mathcal{C}_{<3}$ be defined in $\mathcal{C}_1$ by $f((r_0)) = (0)$ and in $\mathcal{C}_2$ as $f((r_0; r_1)) = (r_1; 0)$. Then, we can see that f is non coherent for each real number. If we look at, say, 1, then, $f((1)) = (0)$ but, for example, $f((1;3)) = (3;0)$. Clearly $(1;3) \in \mathcal{M}_{(1)}$ but $f((1;3)) \notin \mathcal{M}_{f((1))} = \mathcal{M}_{(0)}$.

Definition 2.14 considers a function defined on $\mathcal{C}_{<\beta}$, and imposes a condition on the different levels of definition for $\alpha < \beta$. We have a related notion, this time for functions defined just at one level $\mathcal{C}_\beta$. A function is *stratified* if its image at the α-th component (for $\alpha < \beta$) doesn't depend on finer infinitesimal levels in the domain.

Definition 2.15. $f : \mathcal{C}_\beta \to \mathcal{C}_{Ord}$ is stratified iff it can be written in the form:

$$f(s) = f((s_0; s_1; \ldots; s_\alpha; \ldots)) = (f_0(s_0); f_1(s_0; s_1) \ldots; f_\alpha(s_\lambda)_{\lambda \leq \alpha}; \ldots).$$

Example 2.5. Consider the f of Example 2.4, but now defined only on $\mathcal{C}_2$. Then $f((r_0; r_1)) = (r_1; 0)$, so f_0 depends on r_1 and not only on r_0, so $f|\mathcal{C}_2$ is not stratified.

As illustrated by Examples 2.4 and 2.5, the notions of coherence and stratification are related:

Theorem 2.8. If a function $f : \mathcal{C}_{<\beta} \to \mathcal{C}_{Ord}$ is coherent then each restriction $f|\mathcal{C}_\alpha$ for $\alpha < \beta$ is stratified.

Proof. If $f|\mathcal{C}_\alpha$ for some $\alpha < \beta$ is not stratified then there are $\mu, \nu \leq \alpha$ such that $\mu < \nu$ and the μ-th component of f depends on the ν-th. This means that at least for an $s \in \mathcal{C}_\mu$ there are two sequences $s', s'' \in \mathcal{M}_s$ of length ν such that $f_\mu(s') \neq f_\mu(s'')$. Given this, $f_\mu(s')$ and $f_\mu(s'')$ cannot both coincide with the μ-th value of $f(s)$. Therefore we cannot have that $f_\mu(s')Ef(s)$ and $f_\mu(s'')Ef(s)$ at the same time, even if $s', s'' \in \mathcal{M}_s$, so coherence fails. □

Definitions 2.14 and 2.15 do not capture the intuitive idea of continuity, allowing functions like the Dirichlet-style $^*f : \mathcal{C}_2 \to \mathcal{C}_2$ of Example 2.1. This function is stratified and coherent with f, but is intuitively highly discontinuous. The next definition proposes different levels of continuity using the $\epsilon - \delta$ criterion:

Definition 2.16. Given ordinals $\alpha < \beta$, the function $f : \mathcal{C}_{<\beta} \to \mathcal{C}_{Ord}$ is α-continuous in $s \in \mathcal{C}_\alpha$ if for every positive $\epsilon \in \mathcal{C}_\alpha$ there is a positive $\delta \in \mathcal{C}_\alpha$ such that if $|s - t| < \delta$ then $|f(s) - f(t)| < \epsilon$. f will be said to be α-continuous iff it is α-continuous for every $s \in \mathcal{C}_\alpha$. f will be said to be continuous iff it is α-continuous for every $\alpha < \beta$.

From the above definitions it follows that every continuous function is coherent.

2.9.3 Micro-straightness

Close to this principle of continuity underlying Leibniz's intuitions on the Calculus, is the principle of micro-straightness which states that curves are constituted by straight lines at an infinitesimal level. Barrow talked about "linelets" and "timelets" to indicate these infinitesimal micro segments used for instance by de L'Hôpital.[93] As mentioned before, in SIA we obtain this kind of behavior which is used to develop calculus, something that will be done next also in the present framework. In our context, in fact, we can have different levels of micro-straightness:

Definition 2.17. Consider a coherent function $f : \mathcal{C}_{<\beta} \to \mathcal{C}_{Ord}$ for some ordinal $\beta \geq 3$. For an ordinal α such that $\alpha + 1 < \beta$ we will say that f is micro-straight at the level $\alpha + 1$ (or $(\alpha + 1)$-MS) if the $\alpha + 1$-th component of the function $\tilde{f} = f|_{\mathcal{C}_{\alpha+1}} : \mathcal{C}_{\alpha+1} \to \mathcal{C}_{Ord}$ is linear as a function of the $\alpha + 1$-th variable, namely, r_α. In symbols, if

$$\tilde{f}((r_0; \ldots; r_\alpha)) = (\tilde{f}_0((r_0)); \ldots; \tilde{f}_\alpha((r_0; \ldots; r_\alpha); \ldots)$$

then

$$\tilde{f}_\alpha((r_0; \ldots; r_\alpha)) = a r_\alpha + b$$

with a and b real numbers dependent only on $r_0, \ldots, r_y, \ldots$ for $y < \alpha$.

In practice, we will deal with coherent functions, so it is possible for a successor ordinal $\beta = y + 1$ to consider a function f as being defined on $\mathcal{C}_{<\beta}$, or to consider a corresponding $\tilde{f} : \mathcal{C}_y \to \mathcal{C}_{Ord}$ from which it can be reconstructed. Therefore we can apply the notion of micro-straightness to functions defined on a fixed $\mathcal{C}_y$. In the following subsections, the emphasis will be put on micro-straightness at level 2 (or 2-MS), which will suffice for reconstructing the basic results of Calculus for elementary functions.

2.9.4 Polynomial Functions

As a first step, let us consider polynomial functions. By the definition of the sum and the product in the previous section we obtain micro-straightness at the level of $\mathcal{C}_2$.

Lemma 2.1. Let $f(x) = x^n$ considered as a function from $\mathcal{C}_2$ to $\mathcal{C}_{Ord}$. Then f is micro-straight at the level of $\mathcal{C}_2$.

---

[93] See Bell 1998.

Proof. By induction on n. If $n = 0$ then $f(x)$ is the constant function $y = 1$. At the level of $\mathcal{C}_2$, $f((r_1; r_2)) = (1; 0)$, so in the monad of r_1 the function is the null constant function.

If $n = 1$ then $f(x)$ is a straight line. Let $(r_0; r_1) \in \mathcal{C}_2$. Then, in the monad of r_0, the function is again linear, for $f((r_0; r_1)) = (r_0; r_1)$.

Let us suppose that $f(x) = x^n$ is micro-straight, so $f((r_0; r_1)) = (r_0^n; a \cdot r_1 + b; \ldots)$. Then, if $g(x) = x^{n+1}$, then $g((r_0; r_1)) = f((r_0; r_1)) \odot (r_0; r_1) = (r_0^n; a \cdot r_1 + b; \ldots) \odot (r_0; r_1) = (r_0^{n+1}; a \cdot r_1 \cdot r_0 + b \cdot r_0 + r_0^n \cdot r_1; \ldots) = (r_0^{n+1}; (a \cdot r_0 + r_0^n) \cdot r_1 + b \cdot r_0; \ldots)$. When r_0 is fixed, the second component is linear in r_1. □

Since micro-straightness still holds when multiplying x^n by a real number, and also when adding two monomials, we obtain the following:

Theorem 2.9. Let $f(x) = p(x)$ be a polynomial function (with coefficients in $\mathbb{R}$) considered as defined from $\mathcal{C}_2$ to $\mathcal{C}_{Ord}$. Then f is micro-straight at the level of $\mathcal{C}_2$.

Example 2.6. Let us consider the micro-behavior of the function $f(x) = x^3 - 2x + 5$ inside the real number 2. Let us take $(2; r_1)E(2)$. Since $f(2) = 9$ then $f((2; r_1))E(9)$. Also

$$f((2; r_1)) = (2; r_1) \odot (2; r_1) \odot (2; r_1) - 2 \odot (2; r_1) + 5 = (4; 4r_1; r_1^2) \odot (2; r_1) - (4; 2r_1) + 5 =$$

$$= (8; 12r_1; 6r_1^2; r_1^3) + (1; -2r_1) = (9; 10r_1; 6r_1^2; r_1^3).$$

Thus, we see that the function is linear in the second coordinate.

We have the more general result (here without proof) that polynomial functions are micro-straight when considered at the micro-level n for any natural number.

Theorem 2.10. Let $f(x) = p(x)$ be a polynomial function considered as defined from $\mathcal{C}_n$ to $\mathcal{C}_{Ord}$. Then f is micro-straight at the level of $\mathcal{C}_n$.

Let us see that this result holds at the level of $\mathcal{C}_3$ with the function in the previous example in the monad of $(2; 2)$. So, let us consider a $\mathcal{C}_3$ generic element $(2; 2; r_2)$ and calculate its image up to the component 3:

$$f((2; 2; r_2)) = (2; 2; r_2) \odot (2; 2; r_2) \odot (2; 2; r_2) - 2 \odot (2; 2; r_2) + 5 =$$

$$= (4; 8; 4r_2 + 4; 4r_2; r_2^2) \odot (2; 2; r_2) + (-4; -4; -2r_2) + 5 =$$

$$= (8; 24; 12r_2 + 24; \ldots) + (1; -4; -2r_2) =$$

$$= (9; 20; 10r_2 + 24; \ldots).$$

So in the place $((2;2),(9;20)) \in \mathcal{C}_2 \times \mathcal{C}_2$, the linear mapping $r_2 \mapsto 10r_2 + 24$ corresponds to function f.

2.9.5 Preservation Theorems

We come now to preservation of micro-straightness by sums, products and composition. The following is straightforward:

Theorem 2.11. Let $f : \mathcal{C}_{<\beta} \to \mathcal{C}_{Ord}$ and $g : \mathcal{C}_{<\beta} \to \mathcal{C}_{Ord}$ be both $(\alpha+1)$-MS. Then $f+g$ is also $(\alpha + 1)$-MS.

Considering the product, we have also preservation at the level of $\mathcal{C}_2$:

Theorem 2.12. Let $f : \mathcal{C}_\beta \to \mathcal{C}_{Ord}$ and $g : \mathcal{C}_\beta \to \mathcal{C}_{Ord}$ be both stratified and 2-MS. Then $f \odot g$ is also 2-MS.

Proof. Since f and g are stratified, we have that they can be written in the form:

$$f((r_0;r_1;\dots)) = (f_0((r_0));f_1((r_0;r_1));\dots)$$

and

$$g((r_0;r_1;\dots)) = (g_0((r_0));g_1((r_0;r_1));\dots).$$

Since these functions are 2-MS, then

$$f_1((r_0;r_1)) = ar_1 + b$$

and

$$g_1((r_0;r_1)) = cr_1 + d$$

where a, b, c, d are constant for a given r_0.
We have that

$$(f \odot g)((r_0;r_1;\dots)) = f((r_0;r_1;\dots)) \odot g((r_0;r_1;\dots)) =$$
$$= (f_0((r_0)) \cdot g_0((r_0)); f_0((r_0)) \cdot (cr_1 + d) + g_0((r_0)) \cdot (ar_1 + b);\dots) =$$
$$= (f_0((r_0)) \cdot g_0((r_0)); (cf_0((r_0)) + ag_0((r_0)))r_1 + df_0((r_0)) + bg_0((r_0));\dots).$$

The second component is of the form Ar_1+B, so we obtain the 2-micro-straightness of $f \odot g$. □

Theorem 2.13. Let $f : \mathcal{C}_\beta \to \mathcal{C}_{Ord}$ and $g : \mathcal{C}_\gamma \to \mathcal{C}_{Ord}$ be both stratified and $(\alpha + 1)$-MS. If they are composable then $f \circ g$ is also $(\alpha + 1)$-MS.

Proof. Since g is $(\alpha + 1)$-MS, the $(\alpha + 1)$-th component of $g((r_0;..;r_\alpha;...))$ is of the form $g_\alpha(r_1;..;r_\alpha) = a_g r_\alpha + b_g$ where a_g and b_g depend only on the values of $r_1, ..., r_\lambda...$ for $\lambda < \alpha$. Similarly, since f is $\alpha + 1$-MS, the $\alpha + 1$-th component of $f((r_0;..;r_\alpha;...))$ is of the form $f_\alpha(r_0;..;r_\alpha) = a_f r_\alpha + b_f$ with a_f and b_f dependent also only on $r_1, ..., r_\lambda...$ for $\lambda < \alpha$.

Take $(r_0;..;r_\alpha;...) \in \mathcal{C}_\beta$ such that $g((r_0;..;r_\alpha;...))$ is in the domain of f. Then the $\alpha + 1$-th component of $f \circ g((r_1;..;r_\alpha;...))$ is of the form

$$(f \circ g)_\alpha(r_0;..;r_\alpha) = a_f(a_g r_\alpha + b_g) + b_f =$$
$$= a_f a_g r_\alpha + a_f b_g + b_f.$$

This is of the form $Ar_\alpha + B$, where A and B depend only on r_δ with $\lambda < \alpha$, so $f \circ g$ is also $(\alpha + 1)$-MS. □

Theorem 2.14. Let $f : \mathcal{C}_\beta \to cod(f) \subseteq \mathcal{C}_{Ord}$ be $(\alpha + 1)$-MS, stratified and invertible. Let $f^{-1} : cod(f) \to \mathcal{C}_\beta$. Then the function f^{-1} is $(\alpha + 1)$-MS.

Proof. For simplicity, let us make the proof for the case of 2-micro-straightness, which can be generalized for higher ordinals. If

$$f((r_0; r_1;...)) = (f_0((r_0)); f_1((r_0; r_1));...)$$

then

$$f^{-1}((r_0; r_1;...)) = (f_0^{-1}((r_0)); f_1^{-1}((r_0; r_1));...).$$

Since $f_1((r_0; r_1)) = ar_1 + b$ and the function is invertible, then $f^{-1}((r_0; r_1;...)) = \frac{1}{a}r_1 - \frac{b}{a}$, which is linear in r_1, so f^{-1} is 2-MS. □

2.9.6 Rational Functions

Let us consider $f(x) = 1/x$ as a real-valued function. The aim here is to extend this function to infinitesimal levels and to verify that we have 2-micro-straigthness.

Since we limit ourselves here to the level of resolution of $\mathcal{C}_2$, given $(r_0; r_1) \in \mathcal{C}_2$ we need to find a $(r'_0; r'_1)$ such that $(r_0; r_1) \odot (r'_0; r'_1) = (1; 0; ...)$. In order for this to hold, we need that $r_0 r'_0 = 1$ and that $r_0 r'_1 + r_1 r'_0 = 0$. So, we define $^*f : \mathcal{C}_2 \to \mathcal{C}_2$ through the mapping $(r_0; r_1) \mapsto (\frac{1}{r_1}; -\frac{r_1}{r_0^2})$. This function is linear in r_1, so it is 2-MS.

In general, we obtain the function $g(x) = x^{-n}$ as the iterated product of $^*f$. So, for instance,

$$(^*fx)^2 = \left(\frac{1}{r_0^2}; -\frac{2r_1}{r_0^3}; ...\right)$$

$$(^*fx)^3 = \left(\frac{1}{r_0^3}; -\frac{3r_1}{r_0^4}; ...\right).$$

Here, again, we obtain micro-straightness at the level of $\mathcal{C}_2$ since the second component is linear in r_1.

As a consequence we obtain the following :

Theorem 2.15. Let $f(x) = \frac{p(x)}{q(x)}$ be a rational function considered as defined from $\mathcal{C}_2$ to $\mathcal{C}_{Ord}$. Then f is micro-straight at the level of $\mathcal{C}_2$.

Proof. $\frac{1}{q(x)}$ is 2-MS by Theorems 2.10 and 2.13. We apply then Theorem 2.12 to $f(x) = p(x) \odot \frac{1}{q(x)}$. □

2.9.7 Irrational Functions

Functions like $y = \sqrt{x}$ may be defined on $\mathcal{C}_2$ and shown to be 2-MS following the schema of Theorem 2.14.

Example 2.7. Consider the real-valued function $f(x) = x^2$ and using its formula extend it to $^*f : \mathcal{C}_2^{>0} \to cod(^*f) \subseteq \mathcal{C}_{Ord}$. By Theorem 2.10 f is 2-MS. In fact, $^*f(r_0; r_1) = (r_0^2; 2r_0 \cdot r_1; r_1^2)$. We can then approximate this value as an element of $\mathcal{C}_2$ obtaining $\widetilde{^*f} : \mathcal{C}_2^{>0} \to cod(\widetilde{^*f}) \subseteq \mathcal{C}_2$. Take the inverse:

$$(\widetilde{^*f})^{-1} : \mathcal{C}_2^{>0} = cod(\widetilde{^*f}) \to \mathcal{C}_2.$$

This function is 2-MS. In fact, it may be defined on $(r_0; r_1) \in \mathcal{C}_2$ where $r_0 > 0$, through the explicit expression obtained by taking the inverse in each component:

$$\widetilde{^*f}(r_0; r_1) = (y_0; y_1) = (r_0^2; 2r_0 \cdot r_1).$$

Thus $y_0 = r_0^2$ and $y_1 = 2r_0 \cdot r_1$. So $r_0 = \sqrt{y_0}$ and $r_1 = \frac{y_1}{2r_0} = \frac{y_1}{2\sqrt{y_0}}$. Renaming the variables this gives us the expressions for the inverse function:

$$(\widetilde{^*f})^{-1}(r_0; r_1) = \left(\sqrt{r_0}\, ;\, \frac{r_1}{2\sqrt{r_0}}\right).$$

We can see here that the function is 2-MS. Furthermore, we see from the coefficient of r_1 that the value of the derivative of the inverse function is $(f^{-1})'(x) = \frac{1}{2\sqrt{x}}$, the value obtained through usual calculus techniques.

This procedure may be generalized to other irrational functions like, say, $\frac{1}{\sqrt{x^3-5}}$ or $y = \sqrt[3]{x^2 + 4x + 3}$ using appropriate inverse functions. By Theorems 2.10, 2.11, 2.12, 2.13, and 2.14, we obtain that these functions are 2-MS.

2.9.8 Trigonometric Functions

Let us consider in $\mathbb{R}^2$ the trigonometric circumference, namely, the algebraic curve defined by $x^2 + y^2 = 1$. We can consider then its extension on $(\mathbb{R}^2)^2$ which is 2-MS. It can be in fact obtained as the union of the semicircumferences defined by the irrational functions $y = \sqrt{x^2 + 1}$ and $y = -\sqrt{x^2 + 1}$. We have the function $f(x) = sin(x)$ which may be extended as a function $^*f : \mathcal{C}_2 \to \mathcal{C}_{Ord}$. It is in fact defined at an infinitesimal $(0; r_1) \in \mathcal{C}_2$ by $^*sin((0; r_1)) = (0; r_1)$. This holds by 2-micro-straightness, since an "arch" of length $\varepsilon = (0; r_1)$ is straight and coincides with the length of $^*sin(\varepsilon)$. For similar reasons $^*cos(\varepsilon) = 1$. We have then that[94]

$$^*sin((r_0; r_1)) = {^*sin}((r_0; 0) + (0; r_1)) =$$
$$= {^*sin}((r_0; 0)) \odot {^*cos}((0; r_1)) + {^*cos}((r_0; 0)) \odot {^*sin}((0; r_1)) =$$
$$= sin(r_0) + cos(r_0) \cdot (0; r_1) = sin(r_0) + (0; (cos(r_0)) \cdot r_1) =$$
$$= (sin(r_0); (cos(r_0)) \cdot r_1).$$

The second component $(cos(r_0)) \cdot r_1$ in the last step shows the micro-straightness of the function, and, again, the coefficient of r_1 coincides with the value of the derivative of the original function, as usually obtained through other means. A similar approach may be applied to the other usual trigonometric functions.

---

[94] It may be easily verified that the formula for the sine of the addition of angles still holds in this context.

An alternative extension of $sin(x)$ can be done through power series. In this case the power series $sin(x) = x - \frac{x^3}{3!} + \frac{x^5}{5!} - \ldots$ gives us that

$$sin((r_0; r_1)) = (r_0; r_1) - \frac{(r_0; r_1)^3}{3!} + \frac{(r_0; r_1)^5}{5!} - \ldots =$$

$$(r_0; r_1) - \frac{(r_0^3; 3r_0^2 r_1)}{3!} + \frac{(r_0^5; 5r_0^4 r_1)}{5!} - \ldots =$$

$$(r_0; r_1) - (\frac{r_0^3}{3!}; \frac{3r_0^2 r_1}{3!}) + (\frac{r_0^5}{5!}; \frac{5r_0^4 r_1}{5!}) - \ldots$$

This element of $\mathcal{C}_\omega$ is micro-straight in the second component, and performing the infinite sum we obtain in the second component that the coefficient of the common factor r_1 is:

$$1 - \frac{r_0^2}{2!} + \frac{r_0^4}{4!} - \ldots$$

This is the power series expression of the function $cos(r_0)$.[95]

2.9.9 A Micro-Cancellation Principle

In $\mathcal{C}_2$, admitting nilsquare elements, we do not have a cancellation principle in general. For instance, given $a = (0; 1)$ and $b = (0; 2)$ we have $a \odot a = a \odot b$, but we cannot conclude from this that $a = b$. Nevertheless, we may observe the following:

Theorem 2.16. Let $a, b, e \in \mathbb{R}$, and let $\epsilon = (0; e)$ be an infinitesimal of order 1. If $a \odot \epsilon = b \odot \epsilon$ then $a = b$.

2.9.10 Differential Calculus

For a real-valued function f with an associated $^*f : \mathcal{C}_2 \to \mathcal{C}_{Ord}$ which is 2-MS the derivative $f'(x)$ is defined simply as the slope of the linear function provided by 2-micro-straightness. In other words, if $^*f((r_0; r_1)) = (^*f_0((r_0);$

---

[95] This power series approach, which will not be further pursued here, may be adapted to other analytic functions such as the exponential and logarithmic functions.

$^*f_1((r_0; r_1); \ldots) = (f(r_0); a \cdot r_1 + b; \ldots)$ then $f'(r_0) = a$. For the elementary functions considered in the previous subsections with their associated canonical extensions $^*f$, the derivative thus defined coincides with the usual results obtained in Calculus.

The following increment equation[96] then holds:

$$^*f((r_0; r_1)) = {}^*f(x + \varepsilon) = f(x) + \varepsilon f'(x)$$

where $x = r_0$ and $\varepsilon = (0; r_1)$. Hence,

$$\varepsilon f'(x) = {}^*f(x + \varepsilon) - f(x).$$

From this we can derive the following:

Theorem 2.17. Let $f : \mathcal{C}_\beta \to \mathcal{C}_{Ord}$ and $g : \mathcal{C}_\beta \to \mathcal{C}_{Ord}$ be both 2-MS. Then the differentiation operator is linear, namely:
1. $(f + g)' = f' + g'$
2. $(c \cdot f)' = c \cdot f'$.

Proof. To prove (1) let's apply the increment equation to $f + g$ as follows:

$$\varepsilon(f + g)'(x) = {}^*(f + g)(x + \varepsilon) - (f + g)(x) =$$
$$^*f(x + \varepsilon) + {}^*g(x + \varepsilon) - f(x) - g(x) =$$
$$^*f(x + \varepsilon) - f(x) + {}^*g(x + \varepsilon) - g(x) =$$
$$\varepsilon f'(x) + \varepsilon g'(x).$$

Applying Theorem 2.16 we obtain $(f + g)'(x) = f'(x) + g'(x)$, as required.
The proof of (2) is analogous. □

Let us illustrate now how the rule for differentiating the product of functions holds (Leibniz's Rule).

Example 2.8. Let $f(x) = \sqrt{x}$ and $g(x) = 5x^2 + x$ and consider the product $f(x) \cdot g(x)$. By Example 2.7 we have

$$(\widetilde{^*f})(r_1; r_2) = \left(\sqrt{r_1} ; \frac{r_2}{2\sqrt{r_1}} \right).$$

---

[96] "The fundamental equation of differential calculus", according to Bell, Bell 1998.

Similarly, by straightforward calculation,

$$(\widetilde{^*g})(r_1; r_2) = (5r_1^2 + r_1; 10r_1 \cdot r_2 + r_2).$$

We calculate the product of these two functions:

$$(\widetilde{^*f})(r_1; r_2) \odot (\widetilde{^*g})(r_1; r_2) =$$

$$\left(\sqrt{r_1}; \frac{r_2}{2\sqrt{r_1}}\right) \odot (5r_1^2 + r_1; (10r_1 \cdot r_2 + r_2)) =$$

$$\left(\sqrt{r_1} \cdot (5r_1^2 + r_1); \frac{r_2}{2\sqrt{r_1}} \cdot (5r_1^2 + r_1) + \sqrt{r_1} \cdot (10r_1 \cdot r_2 + r_2)\right) =$$

$$\left(\sqrt{r_1} \cdot (5r_1^2 + r_1); \left(\frac{1}{2\sqrt{r_1}} \cdot (5r_1^2 + r_1) + \sqrt{r_1} \cdot (10r_1 + 1)\right) \cdot r_2\right).$$

We can observe that the second component here is linear in r_2. This procedure leads directly to Leibniz rule:

$$(f \cdot g)'(x) = \frac{1}{2\sqrt{x}} \cdot (5x^2 + x) + \sqrt{x} \cdot (10x + 1) = f'(x) \cdot g(x) + f(x) \cdot g'(x).$$

A generalization of the previous procedure leads to Leibniz's Rule in the large, which I will enunciate without proof for the case of algebraic functions, even if further generalizations are foreseeable.

Theorem 2.18. Let f and g be both algebraic functions (2-MS in their domain). Then $(f \cdot g)' = f' \cdot g + f \cdot g'$.

2.9.11 Integrals

There are different ways in order to define integrals in the present context. One approach uses infinite numbers as defined in *Section 2.5*, similar to the use of Hyperintegers in NSA in order to perform infinite sums of infinitesimals.[97] An alternative approach follows the lines of reasoning of SIA, as in Bell,[98] using microstraightness.

---

97 See Goldblatt 1998.
98 Bell 1998.

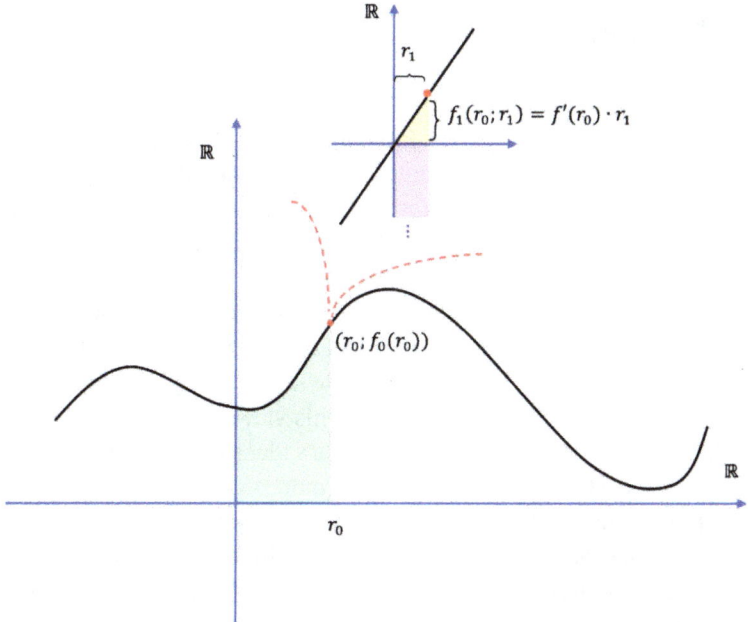

Figure 2.5: In order to calculate the area below the function f up to $(r_0; r_1) = x + \epsilon$ we sum the green area $(A_f(x))$, the violet area $(f_0(r_0) \odot (0; r_1) = f(x) \odot \epsilon)$ and the yellow triangle $(\frac{1}{2}(0; r_1) \odot f'(r_0) = \frac{1}{2} \epsilon \odot f'(x)\epsilon)$.

Let $f : \mathbb{R} \to \mathbb{R}$ have an associated 2-MS extension $^*f : \mathcal{C}_2 \to \mathcal{C}_{Ord}$. We have a function $A_f : \mathbb{R} \to \mathbb{R}$ which assigns to every $x \in \mathbb{R}$ the area below f between 0 and x.[99] Following Leibniz's original ideas,[100] we extend this function to infinitesimal levels. As seen in *Figure 2.5*, this leads to:

$$^*A_f((r_0; r_1)) = {}^*A_f(x + \epsilon) = A_f(x) + f(x) \odot \epsilon + \frac{1}{2} \epsilon \odot f'(x)\epsilon.$$

Since the last term $\frac{1}{2} \epsilon \odot f'(x)\epsilon = \frac{1}{2}f'(x) \odot \epsilon^2 = 0$, we can conclude (from the increment equation) $f(x) = (A_f)'(x)$, so we have a version of the Fundamental Theorem of Calculus:

Theorem 2.19. Let f be a 2-MS function. Then $^*A_f$ is also 2-MS and $f(x) = (A_f)'(x)$.

---

99 This may be justified by the usual Riemann definition.
100 See Tall 1979, Tall 1980, Bell 1998, section 2.3.

2.10 Further Paths

In the first part of this chapter we have seen the possibility of modeling, in a set-theoretic context, the idea of a continuum which transcends the concept of set and that is at the same time non-compositional, reflexive, potential and supermultitudinous. The construction provides a "consistency proof" in the same sense that non-Euclidean geometries were shown to be possible through the construction of adequate models within the Euclidean space.

For Peirce, as I have argued, the consistency of this possibility was part of a wider project of a radical revision of our views of reality and thought. For him, the principle of continuity was "the supreme guide in framing philosophical hypotheses".[101] The philosophical consequences of his views need to be explored and inquired, but as a first step the idea that Peirce's ideas on continuity are not consistent, or a mere "castle in the air", is not any more sustainable.

The second part of the chapter goes in the direction of reconciling Peirce's core ideas of the continuum (and other aspects of his thought, as his modal views on logic and metaphysics, or his defense of the use of infinitesimals in analysis) with other mathematical concepts. The different "views" in the sections of the chapter were aimed to provide possible paths of development, more than finished, final accounts of the possibilities offered by the model constructed. In fact, consistently with Peirce's thought, I conceive the model as reversing the usual conception of mathematical "objects" as completely determined (through analysis which lead to ultimate components) rather than potential and open-ended. From this perspective, many uses in Geometry and Analysis are foreseeable, as I intended to illustrate. In this sense, fields such as algebraic and differential geometry could be approached in an alternative way based on a fundamentally different conception of the underlying spaces.

The key characteristic property, both in the definition of the model $\mathcal{C}_{Ord}$ and its possible developments in other fields, is the stratified, potential character of the entities involved, being them points,[102] lines, manifolds or functions. In this last case, for instance, layered versions of functions in Analysis offer a treatment conceptually different to other approaches to the subject. In the cases examined, the micro-straight behavior of a function in a layer of second order allows the definition of derivatives. This stratified character also may be, I believe, a strength

---

101 [1931–58] CP 6.101, according to Peirce's program of bringing "modern mathematical exactitude into philosophy, and to apply the ideas of mathematics in philosophy" (cited in [2010] p. xviii).
102 Here in the sense of places or loci, and not in the atomistic sense of indivisible entities.

for uses in Geometry, widely understood. It is possible to envision an integration of pointless geometric entities with adaptations of the Kripke model (*Section 2.4*). This can lead to a "modal geometry"[103] incarnating the interactions between the possible and the generic.[104]

---

[103] Zalamea 2012a.
[104] Sheafs are considered as a natural context for the mediation between local and global or generic properties in mathematics. As has been highlighted in Zalamea 2012b, it is one of the crucial concepts of 20th century mathematics. It also allows the interaction between geometry and logic as shown in Mac Lane and Moerdijk 1992, and Caicedo 1995. It has been noticed that natural connections between sheaves and Peircean ideas may emerge in connection to a construction of a continuum model and the idea of genericity (Zalamea 2012b). Kripke models such as the one proposed above may be seen as an advance in this direction.

Arnold Oostra
3 Intuitionistic and Geometrical Extensions of Peirce's Existential Graphs

Abstract: Existential graphs were invented by Charles Peirce as a diagrammatic representation of logic, but his original system was restricted to the classical logic known at the time. In addition to a review of Peirce's initial graphs, in this chapter we show an extension of these diagrams to intuitionistic logic, and we explore the possibility of drawing existential graphs on nonplanar surfaces.

Keywords: Peirce; existential graphs; intuitionistic logic; surface; Jordan curve

Peirce's original work on existential graphs has enormous potential in mathematical logic, and, possibly, in mathematical education and computer science. In this chapter, we show the extension of existential graphs to intuitionistic logic and, on another line, to surfaces different than the usual plane for drawing the diagrams.

Section 3.1 contains a brief presentation of basic existential graphs, complemented by some recent advances in their study. *Section 3.2* is devoted to intuitionistic existential graphs and includes a complete presentation of intuitionistic logic, the full development of the system of intuitionistic Alpha graphs, and a family of extensions to logics closely related to intuitionistic propositional logic. *Section 3.3* explores the possibilities and limitations of developing existential graphs on nonplanar surfaces, first in a general way and then some case studies on specific surfaces.

3.1 Peirce's Existential Graphs

Existential graphs grew out of Peirce's seminal studies in the logic of relatives, now known as predicate logic. While current first-order logic essentially retains Peirce's algebraic presentation of quantifiers, he himself sought a diagrammatic version, finding in existential graphs "a diagram to illustrate the general course of thought"[1] and considering them his *chef d'œuvre*.[2]

---
1 [1906].
2 See Roberts 1973, p. 110.

Arnold Oostra, Arnold Oostra (1966) is Professor at Universidad del Tolima.

https://doi.org/10.1515/9783110717631-003

The subsystems of existential graphs include Alpha, which corresponds to propositional logic, Beta, which constitutes a diagrammatic version of purely relational first-order logic with equality, and Gamma, which opens the way to modal logics and second-order extensions. This chapter focuses primarily on Alpha graphs, although we give the basic indications about Beta and modal Gamma graphs. In this first section, we introduce the standard presentation of existential graphs, specifying some terminology and notation.

3.1.1 Alpha Graphs

The basic Alpha graphs are a diagrammatic representation of classical propositional logic.

3.1.1.1 A current standard presentation

This part is a presentation of the system of Alpha graphs that is very close to Peirce's original versions but, at the same time, is easily formalizable according to current standards of mathematical rigor.

The components from which the Alpha graphs are built are:
– The plane surface, without border, upon which we draw all graphs, called the *sheet of assertion*;
– Propositions, symbolized by *capital letters*;
– Simple closed curves, called *cuts*.

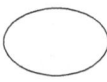

An *Alpha graph* is a diagram composed of a finite combination of letters and cuts, drawn upon the sheet of assertion. There may be repeated letters but they all occupy different places. The cuts do not touch the letters nor do they touch each other. We consider two graphs that can be continuously deformed into each other as equal. This reveals an underlying topology: the graphs can be seen as a part of "topological logic", as we will see profusely in this chapter.

We derive the logical interpretation of the Alpha graphs from the following clauses.
– The sheet of assertion is the universe of possibilities of truth, and drawing a graph on the sheet means asserting its interpretation. Hence, writing a let-

ter means asserting the proposition it represents, and drawing two graphs means asserting both.
- Enclosing a graph with a cut means negating it.

These are the graphs for the basic propositional connectives:
- Negation: ¬A
- Conjunction: A ∧ B   A B
- Disjunction: A ∨ B
- Implication: A → B

From these basic graphs we can build recursively an Alpha graph for any propositional formula. We define that an *area* is a region of the sheet of assertion limited by cuts. An area is *even* or *odd* if there is an even or odd number of cuts around it. A *double cut* is made up of two cuts, one inside the other and without letters or cuts in the area between them.

These are the allowed rules of transformation for Alpha graphs:
1. *Erasure.* In an even area, any graph may be erased.

   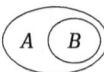   ⇒¹

   A B   ⇒¹   A

2. *Insertion.* In an odd area, any graph may be scribed.

      ⇒²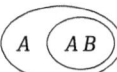

3. *Iteration.* Any graph may be iterated in its own area, or in any area contained in it, that is not part of the graph to be repeated.

      ⇒³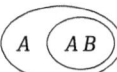

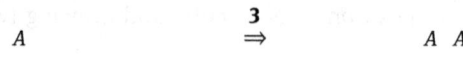

4. *Deiteration.* Any graph may be erased if a copy of it persists in the same area or in any area around it.

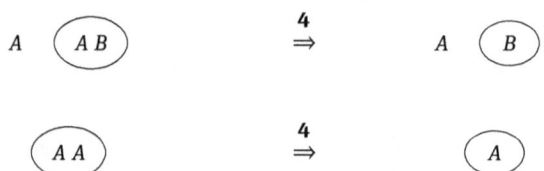

5. *Double cut.* A double cut may be drawn around or removed from any graph on any area.

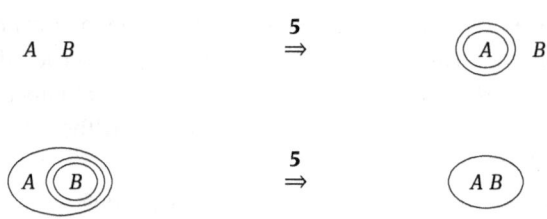

With these rules of transformation, we come to a graphical relation of entailment.

Example 3.1. We can graphically prove *modus ponens*, from A and $A \to B$ follows B:

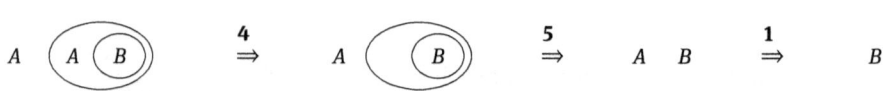

Example 3.2. Without any premise, from the empty sheet of assertion we can prove the law of excluded middle or *tertium non datur*, $A \vee \neg A$:

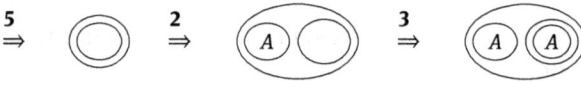

In this way we can graphically prove any result of classical propositional logic. The system of Alpha graphs is a completely equivalent diagrammatic version of this logic.

3.1.1.2 Recent advances

The version of the Alpha graphs presented above condenses Peirce's multiple original presentations, later cleaned up by his scholars in the 1960s. However, we will mention some recent developments regarding this system.

The Alpha decision method and beyond

Peirce proposed a fully graphical method to determine if a given graph is necessary, that is, if it is constructible with the transformation rules starting from the empty sheet of assertion. A slight variant of that method not only decides whether the graph corresponds to a tautology, but indicates its unique place in the free Boolean algebra over the set of propositional letters.

Given any Alpha graph G, we start the procedure with a double cut containing the graph G in its inner area. This area is then separated into two parts by a dotted line, the graph G being entirely in the lower region.

On this graph, we perform the following five operations, which were essentially given by Peirce. Each time, the operation carried out is the first one that is practicable. We will use the sign $\asymp$ to indicate the operations applied. A *simply enclosed letter* is an Alpha graph composed of a cut whose only content is a propositional letter, like $\text{\textcircled{A}}$.

Operation 1. If there is an empty cut, erase the cut that surrounds it directly with all its contents.

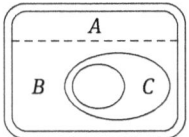 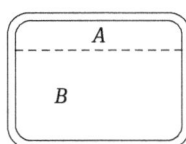

Operation 2. If there is a double cut, erase it leaving its content.

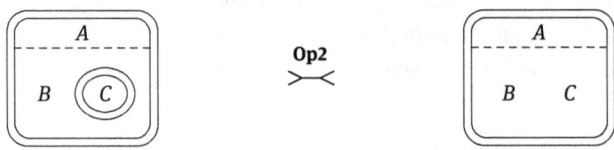

Operation 3. If there is an unenclosed letter in a lower region, write it in the upper region and delete all its occurrences in the lower region.

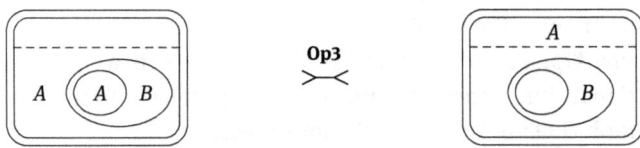

Operation 4. If there is a simply enclosed letter in a lower region, not enclosed by other cuts contained in that region, write it in the upper region, delete all its occurrences in the lower region, and substitute all other occurrences of the letter in that region with empty cuts.

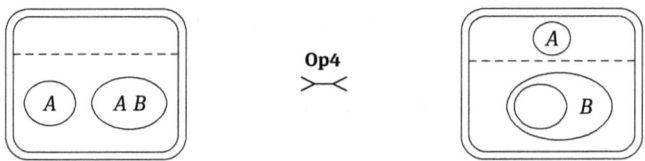

Operation 5. If the above operations cannot be applied to a separated cut and there remain letters in its lower region, then: (*i*) choose any of these letters, (*ii*) iterate the separated cut in its own area with all its contents, (*iii*) in the original cut, write the chosen letter in the upper region and delete all its occurrences in the lower, as in operation 3 and, finally, (*iv*) in the new cut write the chosen letter simply enclosed in the upper region, delete all its simply enclosed occurrences in the lower region, and substitute all its occurrences in that region with empty cuts, as in operation 4.

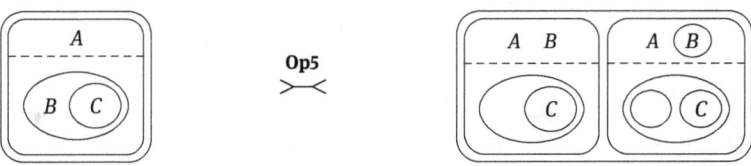

These operations are fully coherent with the Alpha transformation rules. Moreover, in each step, the new graph is equivalent to the given one, and the applied operations provide a standard Alpha proof of this equivalence.[3]

Since the original graph G has a finite amount of letters, this process will end after a finite number of steps. In the end, there are no graphs in the lower regions, and in each upper region there is a combination of different letters and simply enclosed letters. This final graph is a disjunction, and each inner cut contains a combination of letters and negated letters that make the original graph G true.

Example 3.3. Here we apply the method to the graph for the implication $A \to B$. In this case, at first, the only applicable step is operation 5.

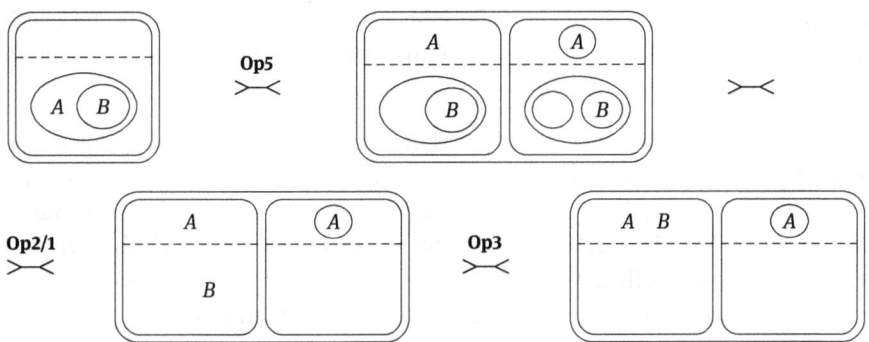

The original graph is true when A and B are both true, or when A is false.

Example 3.4. Let us apply the method to the law of excluded middle $A \vee \neg A$. Again, the only practicable step to start with is operation 5.

---

3 See J. A. Acosta and Jiménez 2010, Díaz 2016, and Oostra 2016.

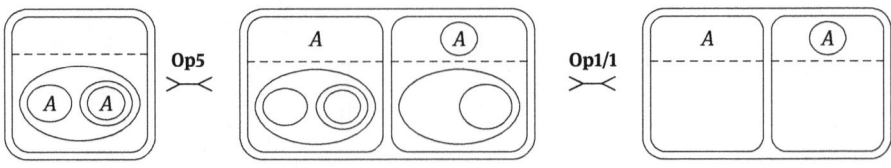

The original graph is true when A is true or when A is false, that is, always. Since all possible combinations of the letter and its negation are present, the graph corresponds to a tautology.

Example 3.5. Finally, we study with this method the contradiction $A \wedge \neg A$. In this case, the first practicable operation is 3.

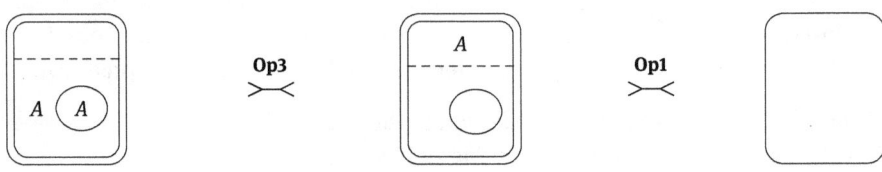

There are no combinations of letters for which this graph is true, hence the graph amounts to a contradiction. Here, the original graph turns out to be equivalent to an empty cut.

We can also use this decision method to determine whether a given graphical reasoning is valid or not. In another line of thought, we can define *graphical normal forms*, and then the decision method guarantees that every Alpha graph is equivalent to a unique graph in full disjunctive normal form.[4] For a set of n propositional letters, there are exactly $2^{(2^n)}$ non-equivalent Alpha graphs in this form. These graphs, partially ordered by graphical entailment, constitute a Boolean algebra whose atoms are all 2^n combinations of different letters and simply enclosed letters. This structure is the free Boolean algebra generated by the set of propositional letters, and is order isomorphic to the Lindenbaum algebra of classical propositional logic.[5]

Formal definition of Alpha graphs

Existential graphs are a diagrammatic version of mathematical logic. In this framework, diagrams are much more than just a mental aid to clarify an idea,

---

[4] See Oostra and Díaz 2016, and Oostra 2016.
[5] Díaz 2016.

without any formal value, as is the case in most mathematical contexts. As in category theory and in *dessins d'enfants*, drawings play a central role here. Some algebraic and order-theoretic definitions of existential graphs have been offered, but that seems a poor substitute for these geometrical and topological objects. A formalization of existential graphs should consider them from the start as true mathematical objects, that is, well defined and susceptible of careful mathematical treatment.

Firstly, we can upgrade the descriptive definition of Alpha graphs given at the beginning of *Subsection 3.1.1.1* to a formal inductive definition, as is customary for logical formulas. However, in these two definitions the components, that is, the sheet of assertion, the letters, and the cuts, are used as primitive notions just as plane, point, and line are used in plane geometry. Secondly, therefore, we can work out precise definitions of these components as mathematical objects. In what follows, we assume the basic notions of general topology without definitions.

For the sheet of assertion, we take the Euclidean plane $\mathbb{R}^2$ with its usual topology. A letter written on the sheet is a point on the plane, labeled with the corresponding propositional letter. Finally, a cut drawn on the sheet is a simple closed curve or *Jordan curve*, which we can more easily consider as the continuous injective image of the unit circle S^1, with its usual topology as a subspace of the plane. To combine different elements, we define an Alpha *pre-graph* as a continuous and injective map $\alpha : mS^1 + F_n \to \mathbb{R}^2$ with an arbitrary labeling function $\lambda : F_n \to L$. Here, m and n are non-negative integers, mS^1 is the topological sum of m copies of the unit circle, $F_n = \{1, 2, \ldots, n\}$ is a finite set with the discrete topology, and L is the set of propositional letters considered. Under these circumstances, α is an embedding, the letters occupy different places, and the cuts do not touch the letters or touch each other.[6] The mapping λ does not need to be injective because there may be repeated letters in a pre-graph.

---

6 Martínez 2014.

Example 3.6. The following pre-graph $\alpha : 2S^1 + F_3 \to \mathbb{R}^2$ draws the premises of *modus ponens*.

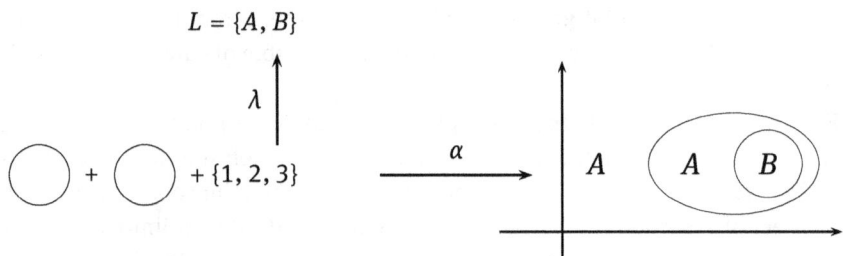

In the first circle, $\alpha(x, y) = (2.5x + 5, 1.5y + 2)$, and, in the second, $\alpha(x, y) = (x + 6, y + 2)$, while $\alpha(1) = (1, 2)$, $\alpha(2) = (3.75, 2)$, and $\alpha(3) = (6, 2)$. On the other hand, $\lambda(1) = \lambda(2) = A$, and $\lambda(3) = B$.

Two pre-alpha graphs $(\alpha : mS^1 + F_n \to \mathbb{R}^2, \lambda)$, $(\alpha' : m'S^1 + F_{n'} \to \mathbb{R}^2, \lambda')$ are *equivalent* if $m = m'$ and $n = n'$ —therefore, with a possible reordering, we can assume that the domains are the same—, also $\lambda = \lambda'$, and, in addition, there exists an isotopy $Y : \alpha \to \alpha'$. An isotopy is a homotopy whose intermediate functions are all embeddings. In this case, it is the same as asking that each intermediate function be also a pre-graph. This relation is compatible with the logical meaning insofar two equivalent pre-alpha graphs share exactly the same logical interpretation.[7] Furthermore, this is an equivalence relation in the set of pre-graphs, and we define an *Alpha graph* as an equivalence class, or isotopy class, of Alpha pre-graphs.

In this way, every Alpha graph is a well-defined mathematical object. On the other hand, the rules of transformation compose a code of permissions to pass from one Alpha graph to another and thus complete an authentic formal system.

Equivalence proofs for Alpha graphs

The proofs of equivalence between Alpha graphs and classical propositional logic have evolved over time. After the first of these arguments, presented by D. Roberts,[8] at least two contemporary proofs have emerged, which are also applicable to systems of existential graphs for some non-classical logics.

---

[7] See Martínez 2014, and also Villareal and Y. Prada 2016.
[8] Roberts 1973, p. 139.

3 Intuitionistic and Geometrical Extensions of Peirce's Existential Graphs

One possible path of proof is the previously mentioned isomorphism between the lattice of Alpha graphs in full disjunctive normal form and the Lindenbaum algebra of classical logic.

A completely novel proof is based upon a new algebraic formal system, that has only letters and parentheses as symbols, and whose rules are inspired by the rules of transformation for Alpha graphs. We represent the conjunction as AB, the negation as (A), and introduce a new constant ⊤ to fill empty areas when required. From these basic conventions, inductively we define the set S_L of *strings* with letters from L. For example, we represent the implication as $(A(B))$, and the disjunction as $((A)(B))$. In the set S_L we select the smallest relation ▷ that satisfies the following clauses.

1. $AB ▷ BA$
2. $A ▷ ⊤$
3. $AB ▷ A$
4. $A ▷ AA$
5. $A(AB) ▷ A(B)$
6. $((A)) ▷ A$
7. $A ▷ ((A))$

Now, a string S *entails* string T, which we denote $S ▶ T$, if there exists a finite sequence of strings $A_1, \ldots, A_n$ with $A_1 = S$, $A_n = T$ and such that $A_i ▷ A_{i+1}$ for each i. In addition, we require the following conditions for ▶.

1. If $A ▶ B$ then $AC ▶ BC$;
2. if $A ▶ B$ then $(B) ▶ (A)$.

In short, ▶ is the smallest transitive relation in S_L that satisfies all nine clauses listed.

With this relation ▶ we can perform all Alpha transformation rules on strings. The meaning of the last two clauses is that if $S ▶ T$, then in any "even" position, we may substitute the string S by T, and T by S in any "odd" position. Therefore, for example, rules 2 and 3 allow us to erase any graph in an even position and, simultaneously, to insert any graph in an odd position. An inductive argument shows that all rules of transformation are completely valid on strings.[9]

---

9 Zambrano 2019.

3.1.2 Beta existential graphs

The heart of Peirce's existential graphs, which were originally intended to represent the relatives or predicate logic, was called system Beta by Peirce himself. We obtain these graphs by adding to the Alpha system a line—which stands for a subject, element, or individual—and letters with attached lines—which stand for relatives, or predicates. To the components of the Alpha graphs we add:
- Thick, possibly branched lines, called *lines of identity*;

- Predicates, symbolized by *capital letters* with a positive number of (ordered) lines attached (the number of lines codifies the arity of the predicate).

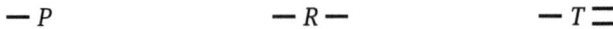

A *Beta graph* is a diagram composed of a finite combination of lines, letters, and cuts drawn upon the sheet of assertion. There may be repeated letters but they all occupy different places. The corresponding number of lines is attached to each letter. If two lines touch then we identify them, but two lines may also cross without touching each other. The cuts do not touch the letters nor do they touch each other, but a line may cross a cut a finite number of times. We consider two graphs that can be continuously deformed into each other as equal.

The interpretation of the Beta graphs follows from the same clauses that we gave for the Alpha graphs, to which we add the following:
- Drawing a line on the sheet means asserting the existence of an individual.
- Writing a letter with lines attached to it means that the predicate it represents holds for the involved individuals.
- Joining two lines of identity means identifying the individuals they represent.

We have the following graphs for the basic quantifiers:
- There exists P: $\qquad \exists x P(x) \qquad$ ——P

- There exists not P: $\qquad \exists x \neg P(x)$

- All is P: $\qquad \forall x P(x)$

- All is not P: $\quad\forall x\neg P(x)$

The square of opposition takes this form in the realm of existential graphs:
- All S are P $\hfill$ No S are P -

$\forall x(S(x) \to P(x))$ $\hfill$ $\forall x(S(x) \to \neg P(x))$

$\exists x(S(x) \wedge P(x))$ $\hfill$ $\exists x(S(x) \wedge \neg P(x))$

- Some S are P $\hfill$ Some S are not P -

The lines of identity make no difference as to the parity of the areas. The rules of transformation that complete the system are the same Alpha rules, only extended with the following adaptations to the line of identity:
1. *Erasure.* In an even area, any line of identity may be cut.
2. *Insertion.* In an odd area, two lines of identity may be joined.
3. *Iteration.* A branch with a loose end may be added to any line; any loose end of a line may be extended inwards through cuts; when there are lines of identity involved in the graph to be iterated, they must correspond exactly to those of the original graph.
4. *Deiteration.* A branch with a loose end may be removed from any line; any loose end of a line may be retracted from the outside through cuts, that is, retracted backwards, reversing what could have been the result of iteration; when there are lines of identity involved in the graph to be deiterated, they must correspond exactly to those of the outside copy of the graph.
5. *Double cut.* The application of this rule is not prevented by the presence of lines that cross both cuts, that is, that pass from outside the outer cut to inside the inner one.

Thus we arrive at a diagrammatic version of first-order logic.

Example 3.7. Follows a graphical proof of syllogism *darii*: All M are P, and some S are M, thus some S are P.

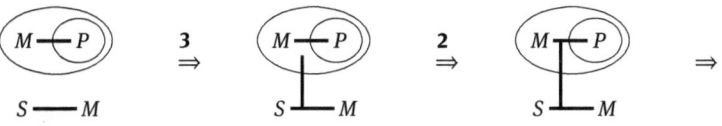

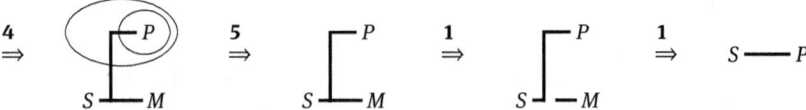

Existential graphs provide a seemingly new classification of the valid syllogisms of Aristotelian logic.[10] On the other hand, the usefulness of Beta existential graphs extends far beyond syllogisms. For example, the elementary theory of binary relations can be developed with existential graphs,[11] as well as basic Euclidean geometry.[12]

3.1.3 Peirce–Zeman's modal Gamma graphs

More than a precise system, like Alpha and Beta, Gamma existential graphs comprise a whole environment where we may introduce and explore new signs. Just one of Peirce's ideas is that of the *broken cut*, a cut drawn with a dashed or discontinuous curve.

Among other options, a feasible interpretation of a broken cut is the *possible negation* of its interior, leading to a representation of alethic modal logic. The combination with continuous cuts brings forth graphs for the basic modalities of possibility and necessity. The following are four alethic modalities with their representation in existential graphs and their customary logical notation.

---

10 Montealegre 2020.
11 Rueda 2011.
12 See Sowa 2020, and Castro 2021.

– Possibly not A Necessarily not A –

$\Diamond \neg A$ $\Box \neg A$

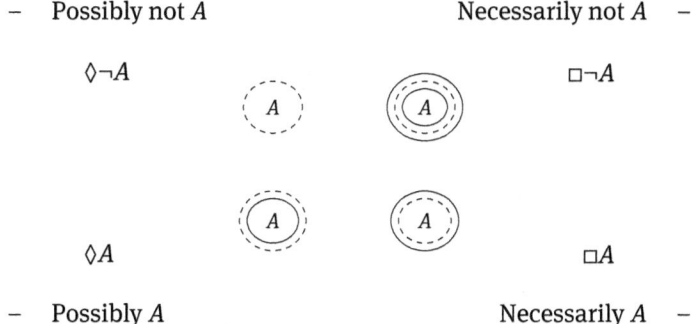

$\Diamond A$ $\Box A$

– Possibly A Necessarily A –

Therefore, a *possible double cut* is a double cut with the outer curve broken and the inner cut continuous, and a *possible graph* is made up of a possible double cut with its contents. The same definitions apply, *mutatis mutandis*, to the necessary case.

To establish the parity of the areas we count the continuous and broken cuts equally. The rules of transformation that complete the modal Gamma systems are all the Alpha rules applied to continuous cuts, and extended with the following adaptations to the broken cuts:

1. *Erasure*. In an even area, a continuous cut may be transformed (by being half erased) into a broken cut.
2. *Insertion*. In an odd area, a broken cut may be transformed (by being filled up) into a continuous cut.
5. *Double cut*. An empty necessary double cut may be drawn or erased in any area.

The first two rules were introduced by Peirce, and allowed him to give one of the first formal proofs in modal logic, see the example below. The third mentioned rule was introduced by J. Zeman, and it is equivalent to the *rule of necessitation* that is usual in current modal logics.

Example 3.8. From $\Box A$ follows A, and from A follows $\Diamond A$.

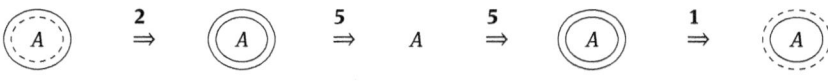

The rule of double cut around any graph is not valid for broken cuts. Neither are the rules of iteration and deiteration, as we show in the following result.

Proposition 3.1. If we may iterate a graph G through broken cuts, then it is equivalent to $\Box G$.

Proof. By Example 3.8, it suffices to show that from G we can obtain $\Box G$.

$$G \quad \overset{5}{\Rightarrow} \quad G\, \overline{()} \quad \overset{3^*}{\Rightarrow} \quad G\, \overline{(G)} \quad \overset{1}{\Rightarrow} \quad \overline{(G)}$$

□

Therefore, if we allow free iteration through broken cuts, this modal system becomes trivial. Different choices of graphs iterable through broken cuts lead to various alethic modal logics. For each selection, the chosen graphs are automatically necessary in that particular system, by Proposition 3.1. Furthermore, the first of the following conditions is no surprise.

- Only necessary graphs may be iterated and deiterated through broken cuts.
- Only necessary or possibly necessary graphs may be iterated and deiterated through broken cuts.
- Only graphs whose minimal components are all inside a broken cut belonging to the same graph, may be iterated and deiterated through broken cuts.

These three instances were introduced by Zeman. In the first case, by Proposition 3.1, from $\Box A$ we obtain $\Box\Box A$, and $\Box A \rightarrow \Box\Box A$ is an axiom characteristic of Lewis' modal logic S4[13] (it is interesting to observe the *natural* emergence of this axiom, called "4" and related to transitivity properties, in Arengas's and Vargas's chapters, *cfr.* pp. 28, 73 above; see also p. 153 below). In fact, the first option gives a system of existential graphs equivalent to S4, the second to S4.2, and the third to S5.

If, in Peirce's words, "in place of a sheet of assertion, we have a book of separate sheets, tacked together at points, if not otherwise connected"[14] then we obtain a blend of Gamma existential graphs and modal Kripke models.[15] This leads to new modal transformation rules that, surprisingly, were already present in Peirce's manuscripts, although they were found after Zeman's thesis.[16] These new rules give place to systems of existential graphs for various other modal logics like Normal, T, and B.

---

[13] Chagrov and Zakharyaschev 1997.
[14] [1931–58] CP 4.512.
[15] R. Prada 2018.
[16] Ma and Pietarinen 2018.

Postscript

At the end of every section in this chapter, we provide a brief commentary with notes on the historical development of the material covered and the main bibliographical references.

Existential graphs were invented by Peirce starting in 1896.[17] Some of his writings on this topic were included in Volume 4 of *Collected Papers*[18] and the first synoptic presentations were given decades later by Roberts,[19] and Zeman.[20] Zeman also developed the modal Gamma graphs and proved their equivalence with different modal logics, and later he studied Peirce's tinctured graphs.[21] More recent presentations of existential graphs include those of P. Thibaud[22] and F. Zalamea.[23] A.-V. Pietarinen recently completed a comprehensive publication of Peirce's manuscripts on existential graphs.[24] There exists at least one book on formal mathematical logic that includes existential graphs.[25]

The Alpha decision method was invented by Peirce and recovered by Roberts,[26] while the mathematical proofs and applications were developed by A. Oostra and his students at the Universidad del Tolima,[27] and he published a summary.[28] The idea of defining Alpha graphs as isotopy classes was introduced by G. Brady and T. Trimble,[29] and improved at the Tolima school.[30] The new equivalence proof between Alpha graphs and propositional logic is based on pioneering ideas by Y. Poveda[31] and was developed by Oostra's students,[32] as was the classification of syllogisms using Beta graphs.[33] Other members of this school inquired

---

[17] Pietarinen 2016.
[18] [1931–58].
[19] Roberts 1963 later published as Roberts 1973.
[20] Zeman 1964.
[21] Zeman 1997.
[22] Thibaud 1975.
[23] Zalamea 2010b and Zalamea 2012a.
[24] [2019–21].
[25] Oostra 2018.
[26] Roberts 1997.
[27] J. A. Acosta and Jiménez 2010, and Díaz 2016.
[28] Oostra 2016.
[29] Brady and Trimble 2000.
[30] Martínez 2014, and also Villareal and Y. Prada 2016.
[31] Poveda 2000.
[32] Taboada and Rodríguez 2010, Fuentes 2014, and Zambrano 2019.
[33] Montealegre 2020.

into Gamma graphs,[34] modal Kripke models,[35] and the connections of existential graphs with category theory.[36] The obstruction result about modal Gamma graphs contained in Proposition 3.1 was discovered by Oostra in correspondence with Zalamea, and published by the former.[37] R. Prada proposed modal Kripke models as sheaves of sheets of assertion.[38] Finally, M. Ma and Pietarinen clarified the new rules in Peirce's manuscripts,[39] and the Tolima school offered a smoother and more complete presentation of systems of Gamma graphs for modal logics.[40]

3.2 Intuitionistic Existential Graphs

In the course of the 20th century, in addition to modal logics, many logics different from the traditional one were formalized, which received the name of *non-classical logics*. Naturally the problem arises to specify existential graphs systems similar to Peirce's for non-classical logics. In this huge program, one of the first logics for which a graphical version is sought is intuitionistic logic. This is due to the close proximity of this logic to topology, and the fact that existential graphs make up a true topological logic.

In this section, we give a complete overview of existential graphs for intuitionistic logic. The first part is a presentation of intuitionistic propositional logic in its customary versions. In the second subsection, we develop the intuitionistic graphs and their transformation rules, and also outline a formal proof of their equivalence with traditional presentations. In the third part, we extend the results to many other logics close to intuitionistic logic.

3.2.1 Intuitionism and intuitionistic logic

Intuitionistic logic is a formal system of symbolic logic developed inside the traditional mathematical realm and that somehow reflects the constructive ideas of intuitionism.

---

34 Molina 2001.
35 López 2013.
36 R. Prada 2012.
37 Oostra 2012, Afirmación 1, p. 31.
38 R. Prada 2018.
39 Ma and Pietarinen 2018.
40 Guerra 2021.

3.2.1.1 A historical note

Intuitionism is a doctrine on the foundations of mathematics that emerged in the first half of the 20th century as a reaction to formalism and logicism. Besides some detached ideas by different thinkers of the previous century, the founder and main pioneer of intuitionism was the Dutch mathematician and philosopher L. E. J. Brouwer (1881–1966). In an intuitionist approach, mathematics is the result of free and constructive mental activity of the person who investigates. In a first move, intuitionism separates mathematics from language and logic and, hence, separates the existence of mathematical objects from their consistence. Furthermore, for Brouwer, the only determining factor of mathematical truth is a mental activity, hence a mathematical proposition becomes true when the subject experiences or *intuits* its truth, after having carried out a suitable mental construction. This idea of truth in turn leads to an intuitionistic interpretation of logical connectives. For example, the negation of a sentence means that something absurd may be constructed from it, and the disjunction of two sentences is true if an effective construction of any of them is possible.

The general ideas of intuitionism lead to logical and mathematical results that differ from the usual ones. For example, the intuitionistic concept of negation compels us to reject the *principle of double negation*. Indeed, the double negation of a proposition means that absurdity can be constructed from the denied sentence, which in no way amounts to an effective construction of the proposition. Very close to this consequence is another which is a well-known feature of intuitionism, the failure of the *principle of excluded middle* or *tertium non datur*. Because, to affirm truth (of a sentence or its denial) in the intuitionistic context, means that one can show an effective construction (of either the sentence or its denial). We may easily state propositions from which no construction can be given and from whose acceptance no contradiction can be built. In classical thought, we acknowledge that one of the two is true, but in the intuitionistic case this is unacceptable *a priori*. Another consequence of the rejection of these principles is the dropping of *reductio ad absurdum*, so beloved in classical mathematics: if we assume the negation of a statement and this leads us to a contradiction, then we can conclude that it is true. Many important results of mathematics are proved by this argument and hence they are no longer valid in intuitionism.

From 1912 on, Brouwer devoted his life to developing a complete revision of mathematics under intuitionistic principles. The result of his effort and that of his successors is a completely different mathematical building than the usual one. While there are many common results, in both theories there are results that are not valid in the other. The purest line of intuitionism has given rise to various species of constructive mathematics that continue to be cultivated to this day, but which do not enter the paths followed by mainstream mathematics.

These general principles of intuitionism also led, within the formal realm of usual mathematics, to the very precise formal system known as *intuitionistic logic*. One of the pioneers of this development was the Dutch mathematician Arend Heyting, a disciple of Brouwer and a tireless promoter of his ideas. In 1930 he published his best-known work, *The Formal Rules of Intuitionistic Logic*.[41] Although Heyting's axiomatization is the most known and cited, there were other important efforts in the formalization of Brouwer's ideas, due to the Russian mathematicians Andrei Kolmogorov and Valerii Glivenko. Afterwards, this logic was worked out further by Gerhard Gentzen and Kurt Gödel, the latter of whom indicated that intuitionistic logic is richer for the simple reason that it distinguishes formulas that are equivalent in classical logic.

Intuitionistic logic, like classical logic, has a wide range of semantic models. A certain class of partially ordered sets, called first *Brouwerian lattices*, are now known as *Heyting algebras* and technically constitute the algebraic counterpart to intuitionistic propositional logic. The most important instance of a Heyting algebra are the open sets of a topological space, ordered by inclusion. On the other hand, Saul Kripke discovered that a certain variant of his models, proposed first as semantics for modal logics, also provided a semantics for Heyting's intuitionistic logic.

In an unexpected turn of the screw, intuitionistic logic appeared naturally and surprisingly in another entirely different mathematical context. A Kripke model for intuitionistic logic may be seen, in particular, as a *sheaf*. Sheaves, originally conceived as tools for algebraic topology, were employed by the French mathematician Alexander Grothendieck and his successors in solving Weil's conjectures in algebraic geometry. At this stage of its development, sheaf theory adopted the language of *categories*. Around 1970 the combination of sheaves with categories gave rise to the theory of *elementary toposes*, a generalized and synthetic environment for mathematics, and very soon it turned out that every topos has an internal logic. Amazingly, this natural logic of topos and sheaves turns out to be intuitionistic. Moreover, with sheaves it is feasible to build models for all intermediate logics between the intuitionistic and classical systems.

---

41 Heyting 1930.

3.2.1.2 Intuitionistic propositional logic

The essential differences between classical and intuitionistic logics lie in the behavior of its connectives, hence we will restrict our whole discussion to the propositional level. In its early presentations, intuitionistic logic had the same logical connectives as classical logic, but later the convenience of including a constant connective representing absurdity was observed.

The basic connectives for intuitionistic propositional logic (IPL) are implication $\to$, conjunction $\wedge$, disjunction $\vee$, and absurd $\bot$. Negation $\neg(\alpha)$ is *defined* as $(\alpha) \to \bot$. The alphabet for IPL is the disjoint union of the sets L of propositional letters, the connectives, and the parentheses. The set $\mathfrak{F}_L$ of formulas is defined inductively, and the unessential parentheses are omitted, all as in the classical case.[42]

The following formulas are taken as axioms of IPL.
1. $\alpha \to (\beta \to \alpha)$
2. $(\alpha \to (\beta \to \gamma)) \to ((\alpha \to \beta) \to (\alpha \to \gamma))$
3. $\bot \to \alpha$
4. $(\alpha \wedge \beta) \to \alpha$
5. $(\alpha \wedge \beta) \to \beta$
6. $(\gamma \to \alpha) \to ((\gamma \to \beta) \to (\gamma \to (\alpha \wedge \beta)))$
7. $\alpha \to (\alpha \vee \beta)$
8. $\beta \to (\alpha \vee \beta)$
9. $(\alpha \to \gamma) \to ((\beta \to \gamma) \to ((\alpha \vee \beta) \to \gamma))$

Axiom 3 stands for the *principle of explosion* or *ex falso quodlibet*: from falsehood, or absurdity, everything follows. The other axioms are exactly the same as for classical propositional logic.

The only inference rule is *modus ponens* (MP): from formulas $\alpha \to \beta$ and α we may proceed to β. A set of formulas Σ *entails* a formula φ, which we denote $\Sigma \vdash \varphi$, if there exists a finite sequence of formulas $\varphi_1, \varphi_2, \ldots, \varphi_n$ (called a *deduction*) with $\varphi_n = \varphi$ and such that every term φ_i has the form of an axiom, or belongs to Σ, or follows from two previous terms by *MP*. A formula τ entailed by the empty set, that is $\vdash \tau$, is called a *theorem* of IPL. Two formulas α, β are *equivalent*, which we denote $\alpha \approx \beta$, if $\alpha \vdash \beta$ and $\beta \vdash \alpha$.

Example 3.9. We may prove the following results just as in classical logic:
- $\alpha \to \beta, \beta \to \gamma \vdash \alpha \to \gamma$;
- $\vdash \alpha \to \alpha$;

---
[42] See, e.g., Caicedo 1990.

- $\alpha, \beta \vdash \alpha \wedge \beta$.

Some remarkable deductions follow at once from the intuitionistic negation.

Example 3.10. $\neg \alpha \vdash \alpha \to \beta$.

Proof. By definition we have $\alpha \to \bot$, and by Axiom 3 we get $\bot \to \beta$. Hence, by the first instance of Example 3.9, we deduce $\alpha \to \beta$. □

Example 3.11 (*Modus tollendo tollens*). $\alpha \to \beta, \neg \beta \vdash \neg \alpha$.

Proof. By definition we have $\alpha \to \beta, \beta \to \bot$, and by Example 3.9 we get $\alpha \to \bot$. □

Combining Example 3.10 in the form $\neg \beta \vdash \beta \to \alpha$ with Axiom 9, as in classical logic, we obtain another time-honoured result.

Example 3.12 (*Modus tollendo ponens*). $\alpha \vee \beta, \neg \beta \vdash \alpha$.

In order to prove an implication in IPL, it suffices to assume the antecedent and deduce the consequent. The proof of this result is the same as in the classical case.[43]

Theorem 3.1 (Deduction theorem). For any set Σ of formulas of IPL:

$$\Sigma \vdash \alpha \to \beta \quad \text{if and only if} \quad \Sigma, \alpha \vdash \beta.$$

Combining this with the third instance of Example 3.9, we may state the following special case of the deduction theorem.

Corollary 3.1. For any formulas α, β, and γ of IPL:

$$\gamma \vdash \alpha \to \beta \quad \text{if and only if} \quad \gamma \wedge \alpha \vdash \beta.$$

In order to prove a negation in IPL, a weak form of *reductio ad absurdum* is available which follows at once from the deduction theorem and the definition of negation as an implication.

Theorem 3.2 (Weak *reductio ad absurdum*). For any set Σ of formulas:

$$\Sigma \vdash \neg \alpha \quad \text{if and only if} \quad \Sigma, \alpha \vdash \bot.$$

One part of *double negation* is valid in intuitionistic logic.

Example 3.13. $\alpha \vdash \neg \neg \alpha$

Proof. By Theorem 3.2 it suffices to show $\alpha, \neg \alpha \vdash \bot$ but by definition this is $\alpha, \alpha \to \bot \vdash \bot$, which follows at once by *MP*. □

---

43 Caicedo 1990, p. 44.

For some formulas *double negation* holds in IPL.

Example 3.14. $\neg\neg\neg\alpha \approx \neg\alpha$

Proof. In one direction, $\neg\alpha \vdash \neg\neg\neg\alpha$ by Example 3.13. In the other way, by Theorem 3.2 it is enough to prove $\neg\neg\neg\alpha, \alpha \vdash \bot$. Again by Example 3.13, from these premises we obtain $\neg\neg\neg\alpha$ and $\neg\neg\alpha$, and as before $\neg\neg\neg\alpha, \neg\neg\alpha \vdash \bot$. □

To end, we may explore some usual relationships between connectives in the intuitionistic case.

Example 3.15. $\neg\alpha \vee \beta \vdash \alpha \to \beta$

Proof. By the deduction theorem (Theorem 3.1) this is equivalent to $\neg\alpha \vee \beta, \alpha \vdash \beta$. But $\alpha \vdash \neg\neg\alpha$, and $\neg\alpha \vee \beta, \neg\neg\alpha \vdash \beta$ by *modus tollendo ponens*. □

Example 3.16. $\alpha \to \beta \vdash \neg(\alpha \wedge \neg\beta)$

Proof. By 3.2 this is the same as $\alpha \to \beta, \alpha \wedge \neg\beta \vdash \bot$, whose proof is straightforward. □

As we might expect from a model for intuitionistic logic, in IPL there is no *double negation*, no *excluded middle*, and no full *reductio ad absurdum*. But even further, although only the axiom of negation was altered, this has an effect on the other connectives since the classical relationships between them do not subsist. In symbols, the following deductions are not possible in IPL.

$$\nvdash \alpha \vee \neg\alpha$$
$$\neg\neg\alpha \nvdash \alpha \qquad\qquad \neg\neg\alpha \not\approx \alpha$$
$$\neg\alpha \to \beta \nvdash \alpha \vee \beta \qquad\qquad \alpha \vee \beta \not\approx \neg\alpha \to \beta$$
$$\alpha \to \beta \nvdash \neg\alpha \vee \beta \qquad\qquad \alpha \to \beta \not\approx \neg\alpha \vee \beta$$
$$\neg(\alpha \wedge \neg\beta) \nvdash \alpha \to \beta \qquad\qquad \alpha \to \beta \not\approx \neg(\alpha \wedge \neg\beta)$$

The only way to ensure that *we can not prove* a certain deduction in IPL requires a semantics for intuitionistic logic, which is the subject of the next part.

3.2.1.3 Semantics for intuitionistic logic

An algebraic semantics for a certain logic is a class of algebraic structures where we may read the formulas and that allows us to decide its validity or truth. In the case of a propositional logic, in order to interpret any formula, we associate the letters with elements of the algebra, which requires operations that translate the different connectives.

An adequate semantics for IPL is given by Heyting algebras, which are a special class of lattices. A *lattice* is a partially ordered set in which every pair of elements a, b has a greatest lower bound, denoted $a \wedge b$, and also a least upper bound, denoted $a \vee b$. A *Heyting algebra* is a lattice H with minimum element $0 \in H$ and a binary operation $\to$ which for any $a, b, x \in H$ satisfies:

$$x \leq a \to b \quad \text{if and only if} \quad x \wedge a \leq b.$$

In other words, $a \to b$ is the greatest element such that $(a \to b) \wedge a \leq b$, an inequality that has a suggestive similarity to *modus ponens*. Notice also that the definition of this connective $\to$ is very reminiscent of Corollary 3.1.

Since $x \wedge a \leq a$ for all elements, the definition implies $x \leq a \to a$, hence every Heyting algebra has a maximum element, which we denote as 1. Moreover, $a \leq b$ if and only if $a \to b = 1$, and in this way the operation $\to$ allow us to express the order relation. On the other hand, the definition implies that every Heyting algebra is a distributive lattice.

In a Heyting algebra we define the *pseudo-complement* $\neg a$ of any element a by

$$\neg a = a \to 0.$$

Again, this is reminiscent of the definition of negation in IPL. Now $x \leq \neg a$ if and only if $x \wedge a = 0$, hence $\neg a$ is the greatest element that is 'disjoint' from a and, in particular, $a \wedge \neg a = 0$. Moreover, we can conclude that $a \leq \neg\neg a$ for every element, and only for some special elements the equality holds, for example $\neg\neg\neg a = \neg a$. We may also establish many other interesting inequalities in Heyting algebras, like $\neg a \vee b \leq a \to b$, $a \vee b \leq \neg a \to b$, and $a \to b \leq \neg(a \wedge \neg b)$.

Follow some examples of Heyting algebras. Any *totally* ordered set with maximum 1 and minimum 0 is a Heyting algebra, if we define:

$$a \to b = \begin{cases} 1 & \text{if } a \leq b; \\ b & \text{otherwise,} \end{cases} \quad \text{thus:} \quad \neg a = a \to 0 = \begin{cases} 1 & \text{if } a = 0; \\ 0 & \text{otherwise.} \end{cases}$$

Hence, for any element a with $0 < a < 1$ we obtain $a \vee \neg a = a \neq 1$, and $a < \neg\neg a$. Moreover, in this case, if $0 < a, b < 1$ then $a \vee b < \neg a \to b$. If $0 < a < b < 1$ then $\neg a \vee b < a \to b$, and if $0 < b < a < 1$, then $a \to b < \neg(a \wedge \neg b)$.

For any positive integer n, the set D_n of all its positive divisors, ordered by divisibility, is a Heyting algebra. In any D_n for $n = p^2q$ with p, q different primes, for instance $n = 12 = 2^2 \times 3$, we may find suitable elements a, b which also satisfy the strict inequalities of the former example.

A *Boolean algebra* is a lattice with maximum 1 and minimum 0 whose binary operations are both distributive over the other, and in which every element a has a unique *complement*, denoted a'. Every Boolean algebra is a Heyting algebra if we define $a \to b = a' \vee b$. The archetypal example of a Boolean algebra is the set of subsets of a fixed universe, ordered by inclusion. The Heyting algebra D_n is a Boolean algebra if and only if n is a square-free integer. In any Boolean algebra, $\neg a = a \to 0 = a' \vee 0 = a'$ and all the following identities hold: $a \vee a' = 1$, $a'' = a$, $a' \to b = a \vee b$, $a \to b = a' \vee b = (a \wedge b')'$. They correspond with the laws of classical propositional logic, in fact we may advance the proportion:

$$\frac{\text{Boolean algebras}}{\text{Classical Logic}} = \frac{\text{Heyting algebras}}{\text{Intuitionistic Logic}}.$$

The archetypal example of a Heyting algebra is the set of *open sets* of a fixed topological space, ordered by inclusion. If U and V are open sets, we define $U \to V = ext(U - V)$, where $ext\, S$ denotes the *exterior* of set S. Hence, the pseudo-complement of any open set U is $\neg U = U \to \emptyset = ext(U - \emptyset) = ext\, U$.

A final example is the Lindenbaum algebra of IPL. The relation $\alpha \approx \beta$ is an equivalence relation in the set $\mathfrak{F}_L$ of all formulas, and the quotient set $\mathcal{L}_L = \mathfrak{F}_L/\approx$ is partially ordered by the (well-defined) relation: $[\alpha] \leq [\beta]$ if $\alpha \vdash \beta$. By the axioms, $[\bot]$ is the minimum element, $[\alpha] \wedge [\beta] = [\alpha \wedge \beta]$, and $[\alpha] \vee [\beta] = [\alpha \vee \beta]$. This lattice is a Heyting algebra since, by Corollary 3.1,

$$[\gamma] \leq [\alpha \to \beta] \quad \text{if and only of} \quad [\gamma] \wedge [\alpha] \leq [\beta],$$

hence $[\alpha] \to [\beta] = [\alpha \to \beta]$. This construction, which consists in naturally assigning an algebraic structure to a given propositional logic, is universally known as a *Lindenbaum algebra*. The maximum 1 of this algebra is the set of all theorems of IPL: $\varphi \in 1$ if and only if $\vdash \varphi$.

In order to establish the Heyting algebras as a semantics for IPL we consider functions or *valuations* $v : L \to H$ from the set of propositional letters to any Heyting algebra H. A valuation v brings about a unique extension function, denoted $\overline{v}$, from the set $\mathfrak{F}_L$ of all the formulas to H. Now a formula φ of IPL *is a consequence* of a set of formulas Σ, and we write $\Sigma \models \varphi$, if for any valuation v such that $\overline{v}(\sigma) = 1$ for each $\sigma \in \Sigma$, we also have $\overline{v}(\varphi) = 1$. A formula τ is *valid* if it is a consequence of the empty set, that is $\models \tau$, if $\overline{v}(\tau) = 1$ for any valuation v.

Example 3.17 (Semantic *modus ponens*). $\alpha \to \beta, \alpha \models \beta$

The following results summarize the resemblance between the entailment and consequence relations, hence between the syntax and semantics of IPL.

Theorem 3.3 (Soundness theorem). For any set Σ of formulas of IPL:

$$\text{If} \quad \Sigma \vdash \varphi \quad \text{then} \quad \Sigma \models \varphi.$$

This is so because all the axioms are valid and *MP* preserves validity, which is the content of the previous example. Hence, any valuation that assigns 1 to the premises also does so to all the formulas of the deduction, in particular to the conclusion. The soundness theorem yields a way to prove that certain deductions *are not possible* in IPL. If there is a suitable valuation v in a Heyting algebra such that $\overline{v}(\sigma) = 1$ for every $\sigma \in \Sigma$ and $\overline{v}(\varphi) \neq 1$, that is to say $\Sigma \not\models \varphi$, then by Theorem 3.3 we conclude $\Sigma \not\vdash \varphi$. Thus, the instances mentioned in specific Heyting algebras show that $\not\vdash p \vee \neg p$, $\neg\neg p \not\vdash p$, $\neg p \to q \not\vdash p \vee q$, $p \to q \not\vdash \neg p \vee q$, and $\neg(p \wedge \neg q) \not\vdash p \to q$.

Theorem 3.4 (Completeness theorem). For any formula φ of IPL:

$$\text{If} \quad \models \varphi \quad \text{then} \quad \vdash \varphi.$$

In particular, the theorems of IPL are exactly the formulas that are valid in Heyting algebras. To prove this result it suffices to define the valuation $v_0 : L \to \mathcal{L}_L$ to the Lindenbaum algebra as $v_0(p) = [p]$, the equivalence class of the letter p. Then $\overline{v_0}(\alpha) = [\alpha]$ for every formula α, and since φ is valid, in particular $\overline{v_0}(\varphi) = [\varphi] = 1$. But then $\varphi \in 1$, which means that $\vdash \varphi$.

A quite different semantics for IPL derives from ordered Kripke models. If P is a set partially ordered by $\leq$, a subset $S \subseteq P$ is *hereditary* if $x \in S$ and $x \leq y$ implies also $y \in S$. Let H_P be the set of all hereditary subsets of P. A *model* based on P is a pair $\mathcal{M} = (P, v)$ where $v : L \to H_P$ is a valuation, which extends to all formulas. For any element $x \in P$, a formula φ is *satisfied in $\mathcal{M}$ at x*, which we denote $\mathcal{M} \models_x \varphi$, if $x \in \overline{v}(\varphi)$. This relation satisfies the following clauses:

- $\mathcal{M} \models_x p$ if $x \in v(p)$;
- $\mathcal{M} \models_x \alpha \wedge \beta$ if $\mathcal{M} \models_x \alpha$ and $\mathcal{M} \models_x \beta$;
- $\mathcal{M} \models_x \alpha \vee \beta$ if $\mathcal{M} \models_x \alpha$ or $\mathcal{M} \models_x \beta$;
- $\mathcal{M} \models_x \alpha \to \beta$ if for all y with $x \leq y$, if $\mathcal{M} \models_y \alpha$ then $\mathcal{M} \models_y \beta$;
- $\mathcal{M} \not\models_x \bot$ for all $x \in P$.

As a consequence, $\mathcal{M} \models_x \neg \alpha$ iff $\mathcal{M} \not\models_y \alpha$ for all $y \geq x$.

A formula φ is *valid in Kripke models* if it is satisfied at all points in any model over every partially ordered set. If τ is a theorem of IPL, that is $\vdash \tau$, then the formula τ is valid in Kripke models. This is another consequence of the soundness theorem (Theorem 3.3), because H_P is a Heyting algebra when ordered by inclusion. Hence, if a certain formula is not satisfied in a given model at some element, then it is not a theorem of IPL. For example, let $P = \{0, 1\}$ with $0 < 1$ as usual and define $v(p) = \{1\}$, that is, for this model $\mathcal{M}$ we define $\mathcal{M} \not\models_0 p$ and $\mathcal{M} \models_1 p$. Since $\mathcal{M} \models_1 p$, we obtain $\mathcal{M} \not\models_0 \neg p$ and therefore $\mathcal{M} \not\models_0 p \vee \neg p$. This is another way to prove that $\not\vdash \alpha \vee \neg \alpha$. Moreover, also $\mathcal{M} \not\models_1 \neg p$ which implies $\mathcal{M} \models_0 \neg\neg p$, and so $\mathcal{M} \not\models_0 \neg\neg p \to p$. Therefore, $\not\vdash \neg\neg\alpha \to \alpha$.

Thus, Kripke models constitute another semantics for IPL.

3.2.2 Intuitionistic Alpha graphs

In this subsection, we develop a Peirce-style system of existential graphs for intuitionistic propositional logic.

3.2.2.1 New signs from old

Although at first the difference between classical and intuitionistic logic lies in the behavior of negation, in fact, this discrepancy goes a long way further. The negative results shown in the previous subsection clearly indicate that in intuitionistic logic we cannot express implication in terms of disjunction and negation, or in terms of conjunction and negation, as we do in classical logic. Certainly, in intuitionistic logic each of the four usual connectives is completely independent of all others, contrary to the strong interconnection that exists between classical connectives. This particular feature of intuitionistic logic forces any graphical representation of it to require different signs for these connectives.

In particular, if we want to apply existential graphics to intuitionistic logic, we might think of changing only the transformation rules. This is what happens with the algebraic presentation of intuitionistic propositional logic, in which we can keep the same formulas and modify only the classical axioms. Clearly, the logical versions of all transformation rules are valid in intuitionistic logic, except for the elimination of a double cut. If we maintain the same Alpha graphs and all other rules, the following graphs are equivalent.

But without the elimination of double cut it seems not feasible to transform this second graph into *B*. In other words, by this way it turns out impossible to obtain *B* from *A* and $A \to B$, that is, we could not even prove *modus ponens*. Which is valid in intuitionistic logic, of course.

Consequently, any system of existential graphs for intuitionistic logic unavoidably needs new signs for implication and disjunction. Our first successful experiments included a diagram for implication in which the inner cut meets the outer cut at one point:

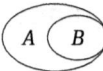

We may also think of it as a continuous curve with one intersection point.

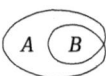

Promptly, Fernando Zalamea pointed out to us that this sign already appears in Peirce's manuscripts. For years Peirce put forward and used, to some extent, alternate diagrams for implication and disjunction in which the inner cuts are connected with the outer one at exactly one point. Here is a sample:[44]

The graph for implication he called a *scroll*, for instance in this outstanding paragraph:

---

44 [1931–58] CP 4.435, 4.457.

Accordingly, since logic has primarily in view argument, and since the conclusiveness of an argument can never be weakened by adding to the premises nor by subtracting from the conclusion, I thought I ought to take the general form of argument as the basal form of composition of signs in my diagrammatization; and this necessarily took the form of a "scroll", that is a curved line without contrary flexure and returning into itself after once crossing itself.[45]

However, Peirce did not imagine that this diagram, taken as a fundamental sign, leads to a system of existential graphs for a more general logic than classical logic. In fact, if we take the scroll as the diagram for implication and the double scroll for disjunction, considering them as new and independent signs, and then apply to the resulting graphs the adapted transformation rules, we obtain a diagrammatic version of intuitionistic logic. That is the content of the next part.

3.2.2.2 The system of intuitionistic Alpha graphs

This part is a simple and plain presentation of a system of existential graphs for intuitionistic propositional logic. We may see this as an extension of Peirce's Alpha system and it is intentionally very similar to it.

I Formation
The components from which the intuitionistic Alpha graphs are built are:
- The plane surface, without border, upon which we draw all graphs, called the *sheet of assertion*;
- Propositions, symbolized by *capital letters*;
- Simple closed curves, called *cuts*;

- Curves called *(single) scrolls*, and composed of two simple closed curves, one of them inside the other and the two intersecting at only one point;

---

45 [1931–58] CP 4.564.

- Curves called *multiple scrolls* and composed of $n+1$ simple closed curves, $n > 1$ of them inside the other one—the inner curves do not touch each other at any point and each one intersects the outer curve at only one, different, point.

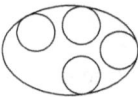

The case of two loops, which we call *double scroll*, deserves special mention.

Later it will become apparent that we could develop the whole system of intuitionistic existential graphs with only single and double scrolls, discarding multiple scrolls with more than two loops.

In all kinds of scrolls, the outer curve is called *cut* and the inner curves *loops*. The region limited by the cut and the loops is the *outer area* of the scroll and the interior of each loop is an *inner area*.

An *intuitionistic Alpha graph* is a diagram composed of a finite combination of letters, cuts, and scrolls, drawn upon the sheet of assertion. There may be repeated letters but they all occupy different places. The cuts and scrolls do not touch the letters nor do they touch each other. We consider two graphs that can be continuously deformed into each other as equal.

At this point we adopt two important conventions. Firstly: *A simple cut enclosing a graph is an abbreviation of a scroll whose loop contains only an empty cut and whose outer area contains only the graph enclosed by the cut.* Thus, for any graph G, we consider the following graphs to be the same.

As we will see shortly, this clause is the graphical version of intuitionistic negation. Peirce anticipated this convention in the classical case:

A scroll with its contents having the pseudograph in the inner close is equivalent to the precise denial of the contents of the outer close.⁴⁶

Secondly: *A multiple scroll with graphs in its areas is an abbreviation of a single scroll whose outer area contains only the same graph of the outer area of the multiple scroll, and whose loop contains only a multiple scroll with an empty outer area and the same loops as the original graph.* For instance, we consider the following graphs to be the same:

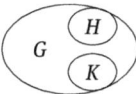   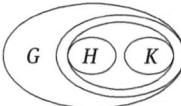

Next, we derive the interpretation of the intuitionistic Alpha graphs from the following clauses.
- The sheet of assertion is the universe of possibilities of truth.
- Drawing a graph on the sheet means asserting its interpretation. Writing a letter means asserting the proposition it represents. Drawing an empty cut means a contradiction.
- Drawing two graphs means asserting both.
- Drawing a scroll means asserting the implication whose antecedent is the graph in the outer area and whose consequent is the graph inside the loop. Hence, by the first of our former agreements, drawing a graph enclosed in a cut without loops means negating it.
- Drawing a multiple scroll with an empty outer area means asserting the disjunction of the graphs enclosed in the different loops. Hence, drawing an arbitrary multiple scroll means asserting the implication whose antecedent is the graph in the outer area and whose consequent is the disjunction of the graphs inside the loops.

Consequently, these are the graphs for the basic intuitionistic connectives:

- Implication: $A \to B$
- Conjunction: $A \wedge B$

---

46 [1931–58] CP 4.456.

- Disjunction: $A \vee B$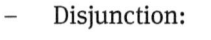

- Absurd: $\bot$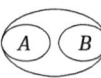

Furthermore, we have the following derived connectives:

- Negation: $\neg A$

- Implication of disjunction: $A \rightarrow (B \vee C)$ 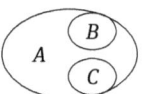 =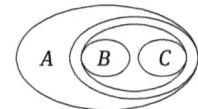

From these basic graphs we can build recursively an intuitionistic Alpha graph for any formula. Now we define that, in general, an *area* is a region of the sheet of assertion limited by curves, both cuts and loops. An area is *even* or *odd* if there is an even or odd number of curves around it, counting both cuts and loops. By this agreement, the following areas are even or odd as we point out:

 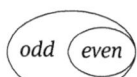 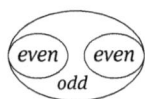

II Transformation

Now we state the rules of transformation allowed for intuitionistic Alpha graphs (basically the same as in the case of classical Alpha graphs, but now applied to the extended context of loops) and illustrate each one with some typical examples.

1. *Erasure*. In an even area, any graph may be erased. Any loop within an even area may be eliminated with its contents.

 $\stackrel{1}{\Rightarrow}$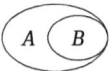

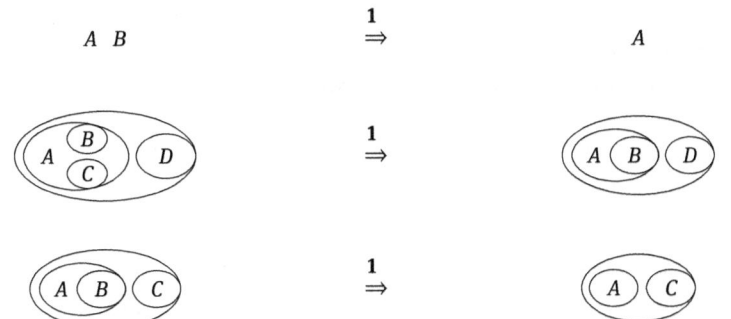

2. *Insertion.* In an odd area, any graph may be scribed. In an odd area limited externally by a cut, a loop containing any graph may be added to this cut.

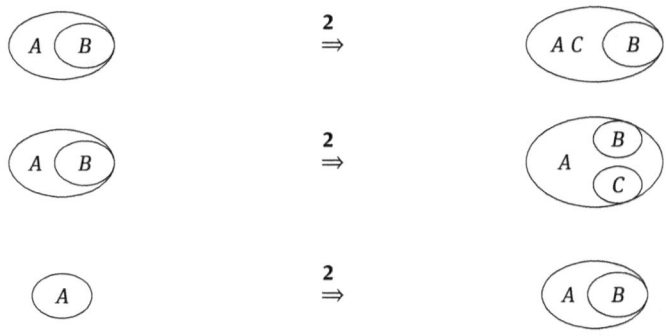

3. *Iteration.* Any graph may be iterated in its own area, or in any area contained in it, which is not part of the graph to be repeated. Any loop may be iterated, with its contents, on its own cut.

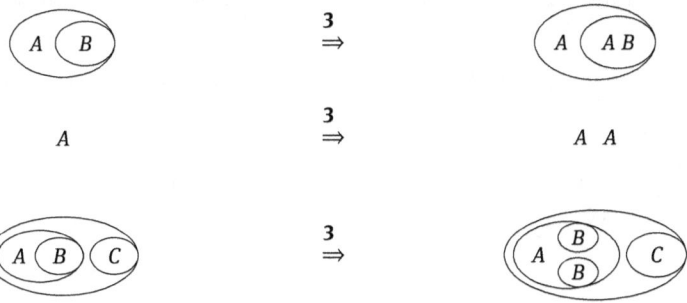

4. *Deiteration.* Any graph may be erased if a copy of it persists in the same area or in any area around it. A loop with its contents may be erased if another loop with the same contents persists on its cut.

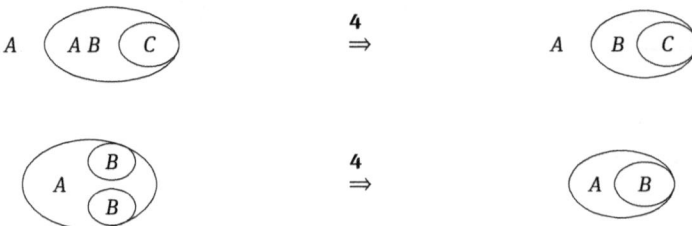

5. *Scrolling.* A scroll with empty outer area may be drawn around or removed from any graph on any area.

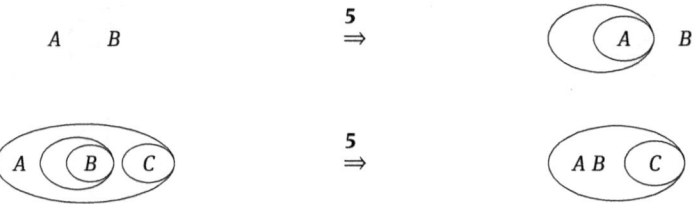

III Deduction

With the given rules of transformation, we come to an intuitionistic graphical entailment. An intuitionistic Alpha graph G entails graph H, which we denote $G \Rightarrow H$, if there exists a finite sequence of Alpha graphs $A_1, A_2, \ldots, A_n$ (called a *deduction*) with $A_1 = G$ and $A_n = H$, such that every graph A_i is obtained from the preceding one A_{i-1} by the application of one of the five rules. In the statement of the entailment we may use the traditional notation for the intuitionistic connectives. Two intuitionistic Alpha graphs G, H are *equivalent*, and we write $G \Leftrightarrow H$, if $G \Rightarrow H$ and $H \Rightarrow G$.

Example 3.18. We can graphically prove *modus ponens*: $A, A \to B \Rightarrow B$.

Graphical statement:

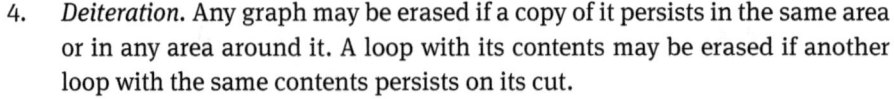

Proof.

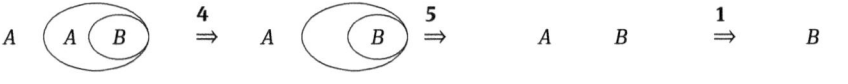

Example 3.19. $A \to B, B \to C \Rightarrow A \to C$

Graphical statement:

Proof.

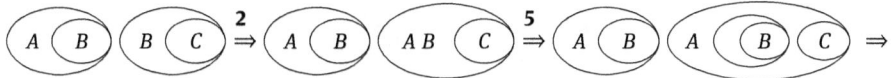

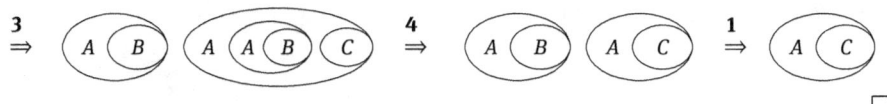

Example 3.20. $\Rightarrow A \to (B \to A)$, without any premise.

Graphical statement:

The only way to start a proof without any premise is by drawing an empty scroll.

Proof.

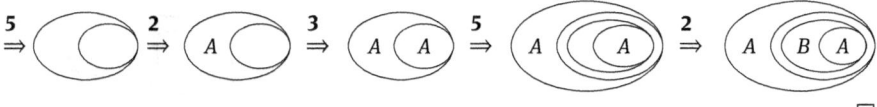

Notice that in the first three steps we graphically prove theorem $A \to A$, see Example 3.9 in *Subsection 3.2.1.2*.

3.2.2.3 Graphical intuitionistic logic

We start with a fully graphical example.

Example 3.21.

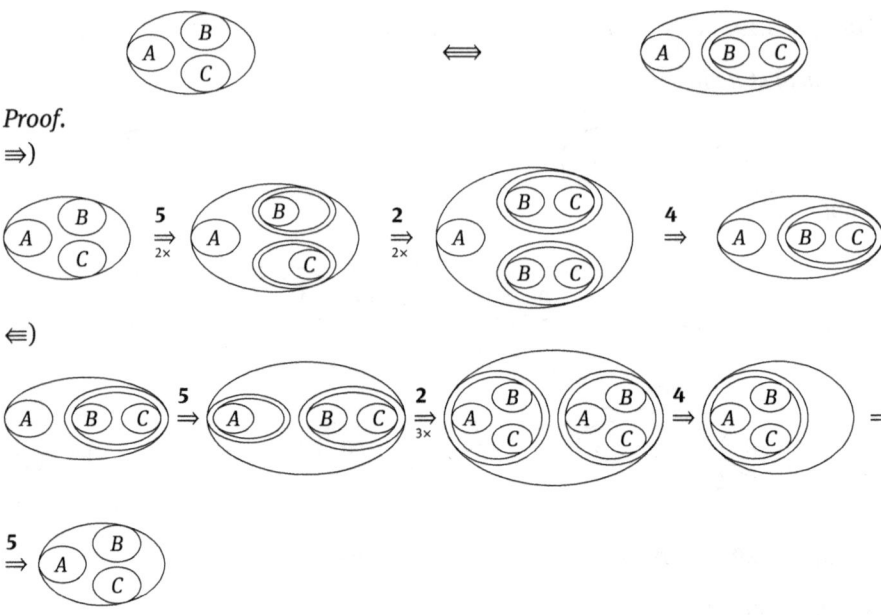

The same result holds for four, five, or more loops. This clearly shows that it is sufficient to consider double scrolls, since any multiple scroll is a composite of them.

Example 3.22. $A \vee (B \vee C) \iff (A \vee B) \vee C$.

Graphical statement:

Proof. This follows from Example 3.21 since, *mutatis mutandis*, both graphs are equivalent to the triple scroll with loops containing A, B, and C. □

We obtain the following graphical deduction also from Example 3.21 by the definitions.

Example 3.23.

Now we will show some graphical proofs corresponding to deductions that we highlighted in *Subsection 3.2.1.2*.

Example 3.24. $A \to B \Rightarrow \neg(A \wedge \neg B)$.

Graphical statement:

Geometrically, this means that a loop within an odd area may be unfastened to form a simple cut. However, it is clear that we will not be able to join it back, because then the intuitionistic scroll would be equivalent to the classical one.

Proof.

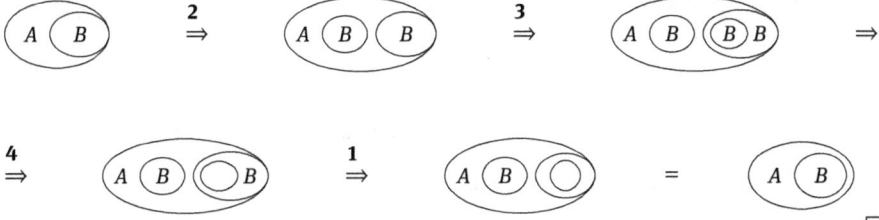

□

Omitting the letter A throughout the entire proof above, and adding a scrolling step, we obtain the next result.

Example 3.25. $B \Rightarrow \neg\neg B$.

Graphical statement:

Also, by adding a deiteration step and an erasure step, as in the classical Alpha graphs, we may derive another important graphical deduction.

Example 3.26 (*Modus tollendo tollens*). $A \to B, \neg B \Rightarrow \neg A$.

Graphical statement:

We may also unfasten the loop of a double scroll, and the same proof applies exactly for multiple scrolls.

Example 3.27. $A \vee B \Rightarrow \neg A \to B$.

Graphical statement:

Proof.

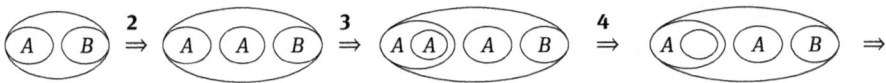

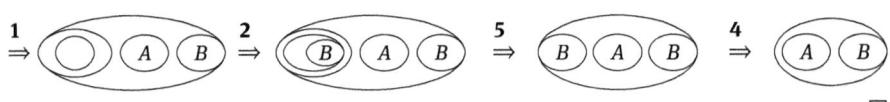

□

From this result, the proof of the following facts is straightforward.

Example 3.28 (*Modus tollendo ponens*).
1. $A \vee B, \neg A \Rightarrow B$;
2. $A \vee B, \neg B \Rightarrow A$.

Graphical statement:

We show one more example of a typical intuitionistic result.

Example 3.29. $\neg A \vee B \Rightarrow A \to B$.

Graphical statement:

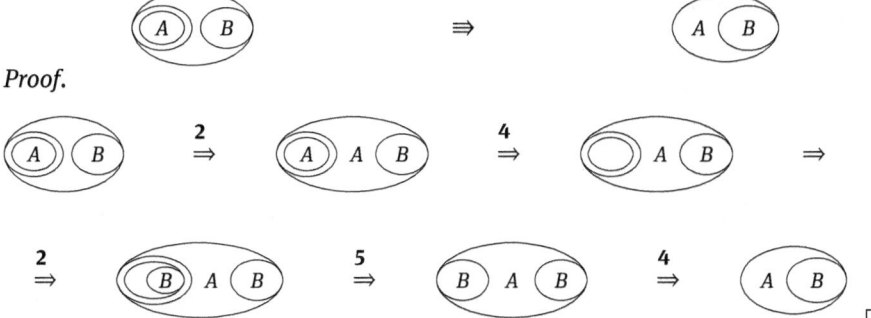

Proof.

We could simplify the last proof somewhat by applying the principle of contraposition, which is also valid in these graphs because of the way the rules are defined.

Theorem 3.5 (The principle of contraposition). *Let G, H be intuitionistic Alpha graphs such that $G \Rightarrow H$.*
1. *Each occurrence of G in an even area of any graph may be transformed into H;*
2. *Each occurrence of H in an odd area of any graph may be transformed into G.*

In the graphical deductions we will indicate with a **C** any step taken with this result.

Proof. It suffices to check all the transformation rules used in the deduction of H from G. For example, if we use the rule of erasure of a graph in some step, if all the steps are carried out again in the even area where G is, then the area for that particular step is still even, and we may perform the same erasure. On the other hand, if all the steps are done in reverse in the odd area where H is, then the area where the erased graph was is now odd, and therefore we may draw it by insertion in odd. And so on for all the five rules and their special cases. □

Example 3.30. $\neg\neg A \to \neg A \Rightarrow \neg A$.

Graphical statement:

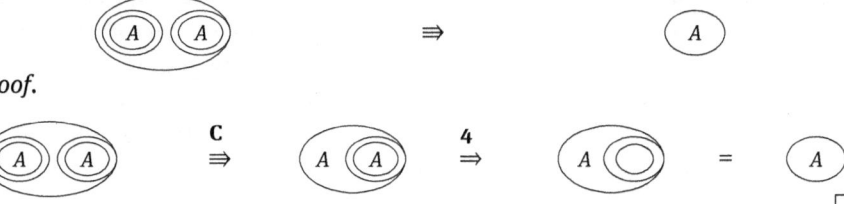

Proof.

Actually, the above proof hides the following result.

Example 3.31. $A \to \neg A \Rightarrow \neg A$.

Graphical statement:

Proof.

$\square$

Is this system of existential graphs really different from classical Alpha graphs? The following result begins to clarify this question.

Proposition 3.2. Let A, B intuitionistic Alpha graphs. If $\neg\neg B \Rightarrow B$ then $\neg(A \land \neg B) \Rightarrow A \to B$.

By Example 3.24, in fact, the conclusion is $A \to B \Longleftrightarrow \neg(A \land \neg B)$.

Graphical statement:

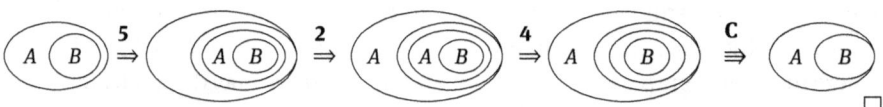

Geometrically, this means that if we assume the transformation rule of double cut, then we might join any simple cut inside another to form a scroll.

Proof.

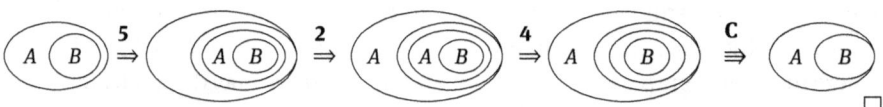

$\square$

It is worth mentioning that, in IPL, if $\neg\neg \beta \vdash \beta$ then $\neg(\alpha \land \neg\beta) \vdash \alpha \to \beta$, and equivalently, in any Heyting algebra, $\neg\neg b \leq b$ implies $\neg(a \land \neg b) = a \to b$.

The true relevance of Proposition 3.2 is that if we admit the elimination of double cuts, then the scroll loses its meaning because it would be equivalent to the classical one, composed of two disconnected cuts. Therefore, this whole new system would become equivalent to the usual classical Alpha graphs. If we assume that the scroll is a novel graph, then the system is not equivalent to the classical Alpha graphs. Furthermore, in the logic corresponding to these graphs the law of double negation is not fully valid, so this system of existential graphs is not classical.

3 Intuitionistic and Geometrical Extensions of Peirce's Existential Graphs

The formal equivalence between IPL and the system of intuitionistic Alpha graphs has many technical details, here we will give only a brief outline.[47]

From the standard drawing of the connectives, we define inductively a mapping g from the set $\mathfrak{F}_L$ of the intuitionistic formulas to the set $\mathcal{G}_L$ of all intuitionistic Alpha graphs on the sheet of assertion with letters from L. The proof that $\alpha \vdash \beta$ implies $g(\alpha) \Rightarrow g(\beta)$ follows from two facts. On the one hand, we can deduce the translation of each axiom of IPL from the empty sheet of assertion, as we did with Axiom 1 as Example 3.20 in *Subsection 3.2.2.2*. On the other hand, *modus ponens* corresponds to a graphical entailment as we showed with Example 3.18 in that same part. In this way, we can remake graphically every step of a deduction for $\alpha \vdash \beta$.

For the other direction of the sought equivalence, which is far harder to prove, we introduce a new formal system for IPL. It has fewer symbols than the usual presentation, and its rules are inspired by the rules of transformation for intuitionistic Alpha graphs. We represent a scroll as $[A(B)]$, a double scroll as $[(A)(B)]$, an empty cut as $[\top]$ with $\top$ a constant not in L, and an arbitrary cut as $[A]$. From these basic conventions, inductively we define the set $\mathcal{S}_L$ of *intuitionistic strings* with letters from L, and a representation function s that assigns an intuitionistic string $s(G)$ to every intuitionistic Alpha graph G. In the set $\mathcal{S}_L$ we select the smallest relation $\triangleright$ that satisfies the following clauses.

1. $AB \triangleright BA$
2. $[(A)(B)] \triangleright [(B)(A)]$
3. $A \triangleright \top$
4. $AB \triangleright A$
5. $A \triangleright AA$
6. $A \triangleright [(A)(B)]$
7. $[\top] \triangleright A$
8. $[AB(C)] \triangleright [AB(BC)]$
9. $A[(B)(C)] \triangleright A[(AB)(C)]$
10. $A[AB(C)] \triangleright A[B(C)]$
11. $[(A)(A)] \triangleright A$
12. $[\top(A)] \triangleright A$
13. $A \triangleright [\top(A)]$

Now, an intuitionistic string S *entails* string T, which we denote $S \blacktriangleright T$, if there exists a finite sequence of strings $A_1, \ldots, A_n$ with $A_1 = S$, $A_n = T$ and such that $A_i \triangleright A_{i+1}$ for each i. In other words, $S \blacktriangleright T$ is the smallest transitive relation on $\mathcal{S}_L$

---

[47] For a complete proof see Ortiz and Segura 2018, and Oostra 2021.

that contains the relation ▷. It is also reflexive by conditions 5 and 4, or 13 and 12. In addition, we require the following conditions for ▶.
1. If $A \triangleright B$ then $AC \triangleright BC$;
2. if $A \triangleright B$ then $[B(C)] \triangleright [A(C)]$;
3. if $A \triangleright B$ then $[C(A)] \triangleright [C(B)]$;
4. if $A \triangleright B$ then $[(A)(C)] \triangleright [(B)(C)]$.

In short, ▶ is the smallest transitive relation on $\mathcal{S}_L$ that satisfies all seventeen clauses. The intuitionistic strings S, T are equivalent, and we write $S \blacklozenge T$, if $S \triangleright T$ and $T \triangleright S$.

With this relation ▶ we can perform all Alpha transformation rules on intuitionistic strings. The meaning of the last four clauses is that if $S \triangleright T$, then in any "even" position, we may substitute the string S by T, and T by S in any "odd" position. Therefore, for example, the first rules 3 and 4 allow us to erase any graph in an even position and, simultaneously, to insert any graph in an odd position. A subtle inductive argument shows that all rules of transformation are completely valid on strings.[48] Thus, for any intuitionistic Alpha graphs G, H, if $G \Rightarrow H$ then $s(G) \triangleright s(H)$.

Finally, we define a mapping f that translates each string back into a formula of IPL. Perhaps, the only cases we need to explicitly specify are $f([\top]) = \bot$ and $f(\top) = \bot \to \bot$. The proof in IPL of the seventeen rules is feasible –easily in most of the cases–, and with this we conclude that if $S \triangleright T$ then $f(S) \vdash f(T)$, for any strings S, T.

The composite fs is not exactly the inverse mapping of g, but we can prove the following:
- $fsg(F) \approx F$ for any formula F of IPL;
- $gfs(G) \Longleftrightarrow G$ for any intuitionistic Alpha graph G;
- $sgf(S) \blacklozenge S$ for any string S.

From this we conclude that the three ordered structures $\mathcal{F}_L/\approx$, $\mathcal{G}_L/\Longleftrightarrow$, and $\mathcal{S}_L/\blacklozenge$ are isomorphic, in fact all are isomorphic to the free Heyting algebra generated by L. Therefore, the system of existential Alpha graphs is equivalent to IPL.

---

48 Ortiz and Segura 2018.

3.2.3 Subsequent extensions

Once a system of existential graphs is established for IPL, the path is open to consider existential graphs for many other non-classical logics associated with intuitionistic logic.

3.2.3.1 Intuitionistic Beta and Gamma graphs

The basic difference between classical and intuitionistic logic lies at the propositional level, thus the essential features of intuitionistic existential graphs are found in the Alpha system. However, it is not difficult to build intuitionistic Beta and modal Gamma graphs.

We obtain the system of intuitionistic Beta graphs by adding to the Alpha system a line—which stands for a subject, element, or individual— and letters with attached lines—which stand for relatives, or predicates—. Formally, to the components of the intuitionistic Alpha graphs we add:
- Thick, possibly branched lines, called *lines of identity*;

- Predicates, symbolized by *capital letters* with a positive number of (ordered) lines attached.

An *intuitionistic Beta graph* is a diagram composed of a finite combination of lines, letters, cuts, and scrolls drawn upon the sheet of assertion. There may be repeated letters but they all occupy different places. The corresponding number of lines is attached to each letter. If two lines touch then we identify them, but two lines may also cross without touching each other (imagine them in three dimensional space, using "Peirce's bridge", that is, with lines going over/under each other). The cuts and scrolls do not touch the letters nor do they touch each other, but a line may cross a cut or loop a finite number of times. We consider two graphs that can be continuously deformed into each other as equal.

The interpretation of the intuitionistic Beta graphs follows from the same clauses that we gave for the intuitionistic Alpha graphs, to which we add the following.
- Drawing a line on the sheet means asserting the existence of an individual.

- Writing a letter with lines attached to it means that the predicate it represents holds for the involved individuals.
- Joining two lines of identity means identifying the individuals they represent.

Therefore, we have the following graphs for the basic intuitionistic quantifiers.

- There exists P: $\quad\exists x P(x)$
- There exists not P: $\quad\exists x \neg P(x)$
- All is P: $\quad\forall x P(x)$
- All is not P: $\quad\forall x \neg P(x)$

In terms of categorical propositions, we have the following graphical representations.

- Some S are P: $\quad\exists x(S(x) \wedge P(x))$
- Some S are not P: $\quad\exists x(S(x) \wedge \neg P(x))$
- All S are P: $\quad\forall x(S(x) \to P(x))$
- All S are not P: $\quad\forall x(S(x) \to \neg P(x))$

Informally, the existential quantifier corresponds to the line and the universal quantifier to the combination of the line with a scroll. As we will prove graphically later, in intuitionistic logic "all S are not P" $\forall x(S(x) \to \neg P(x))$ is equivalent to "no S are P" $\neg \exists x(S(x) \wedge P(x))$, and, in particular, "all is not P" $\forall x \neg P(x)$ is equivalent to "nothing is P" $\neg \exists x P(x)$. On the contrary, however, "all S are P" $\forall x(S(x) \to P(x))$ is *not* equivalent to "it is false that some S are not P" $\neg \exists x(S(x) \wedge \neg P(x))$, and "all is P" is *not* equivalent to "it is false that there exists no P" $\neg \exists x \neg P(x)$.

The lines of identity make no difference as to the parity of the areas. Hence, the rules of transformation that complete the system are the same Alpha rules, only extended with the following adaptations to the line of identity:

1. *Erasure.* In an even area, any line of identity may be cut.

2. *Insertion.* In an odd area, two lines of identity may be joined.
3. *Iteration.* A branch with a loose end may be added to any line; any loose end of a line may be extended inwards through cuts or loops; when there are lines of identity involved in the graph to be iterated, they must correspond exactly to those of the original graph.
4. *Deiteration.* A branch with a loose end may be removed from any line; any loose end of a line may be retracted from the outside (retreated, as a reverse of iteration) through cuts or loops; when there are lines of identity involved in the graph to be deiterated, they must correspond exactly to those of the outside copy of the graph.
5. *Scrolling.* The application of this rule is not prevented by the presence of lines that cross both the cut and the loop of the scroll with empty outer area, that is, that pass from outside the scroll to the inside of the loop.

We give some examples of deductions with intuitionistic Beta existential graphs.

Example 3.32. $\forall (M(x) \to P(x)), \forall x(S(x) \to M(x)) \Rightarrow \forall x(S(x) \to P(x))$.

This is the intuitionistic version of the syllogism *barbara*.

Graphical statement:

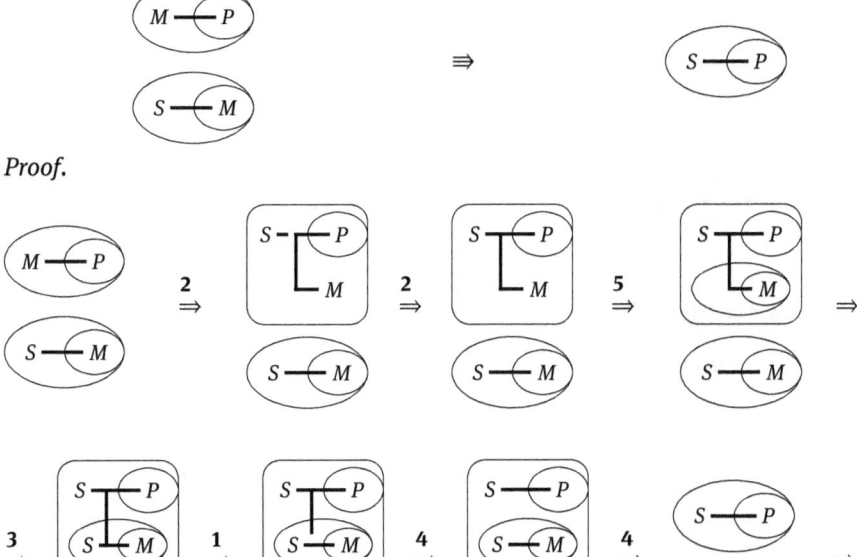

Proof.

$\overset{1}{\Longrightarrow}$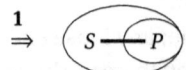

□

Example 3.33. $\forall (S(x) \to \neg P(x)) \iff \neg \exists x (S(x) \land P(x))$.

Graphical statement:

Proof.

$\Longrightarrow$)

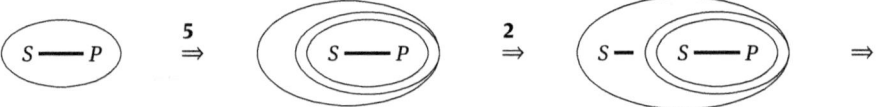

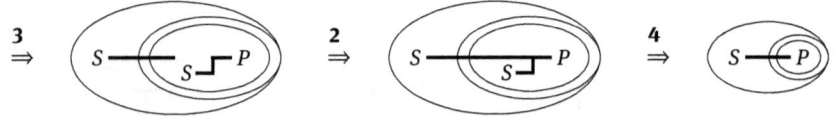

$\Longleftarrow$)

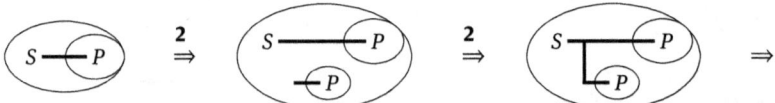

□

Example 3.34. $\forall (S(x) \to P(x)) \implies \neg \exists x (S(x) \land \neg P(x))$.

Graphical statement:

Proof.

3 Intuitionistic and Geometrical Extensions of Peirce's Existential Graphs — 151

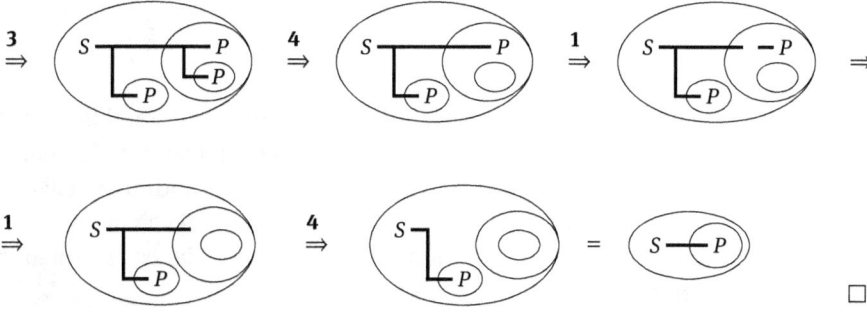

In a similar way, we obtain several systems of intuitionistic modal Gamma graphs by admitting broken lines in the intuitionistic Alpha system. Since both the cut and the loops can be continuous or broken, there are four possible modal scrolls:

Also, there are six possible modal double scrolls:

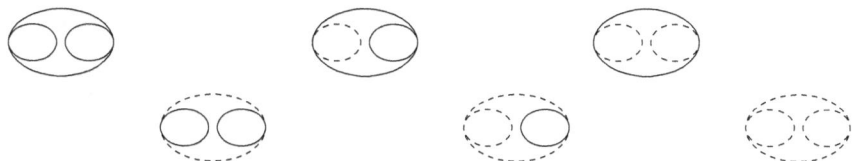

And so on for multiple scrolls. The interpretation of these graphs is the same that we gave for intuitionistic Alpha graphs adding that, in general, a broken cut means that its content is contingent, or not necessary. Here follow some basic intuitionistic modal Gamma graphs with their reading.

- A is necessary: $\quad\quad\quad\quad\quad\quad \Box A$

- A implies that B is necessary: $\quad A \to \Box B$

- A is possible: $\quad\quad\quad\quad\quad\quad \Diamond A$

- It is possible that A implies B: $\quad \Diamond(A \to B)$

- A or it is necessary that B: $\quad A \vee \Box B$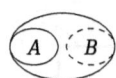

To establish the parity of the areas we count both the continuous and the broken curves. The rules of transformation that complete the intuitionistic modal Gamma systems are again the intuitionistic Alpha rules applied to continuous cuts and loops, and extended with the following adaptations to the broken curves:

1. *Erasure.* In an even area, a continuous cut or loop may be transformed (by being half erased) into a broken curve.
2. *Insertion.* In an odd area, a broken cut or loop may be transformed (by being filled up) into a continuous curve.
5. *Scrolling.* A completely empty scroll with continuous cut and broken loop may be drawn or erased in any area.

Since a scroll with a continuous cut, a broken loop, and an empty outer area symbolizes the necessity of the contents of the loop, this last rule amounts to the modal *rule of necessitation*. With these rules we can prove some basic modal deductions.

Example 3.35. $\Box A \Rightarrow A$.

Graphical statement:

Proof.

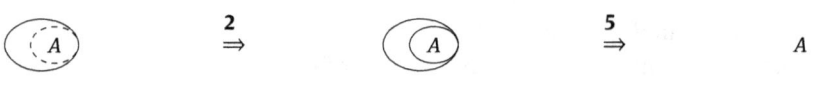

□

As in the classical case, free iteration and deiteration across broken curves has devastating consequences.

Proposition 3.3. If a graph A is iterable across broken curves, then $A \Longleftrightarrow \Box A$.

Proof. It is enough to combine Example 3.35 with the following deduction.

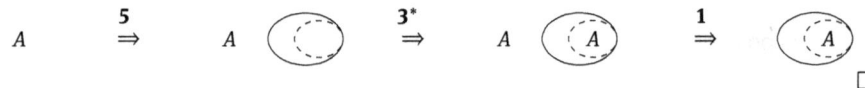

□

This opens up a wide range of possibilities for choosing iterable and deiterable graphs. By Proposition 3.3, the following seems like the minimum option:

3. *Iteration*. The only graphs that may be iterated across broken cuts are the necessary ones, that is, those graphs consisting of a scroll with a continuous cut, a broken loop, and an empty outer area, but any graph within the loop.
4. *Deiteration*. The only graphs that may be deiterated across broken cuts are the necessary ones.

We show a deduction with intuitionistic modal Gamma graphs.

Example 3.36. $\Box A \Rightarrow \Box\Box A$.

Graphical statement:

Proof.

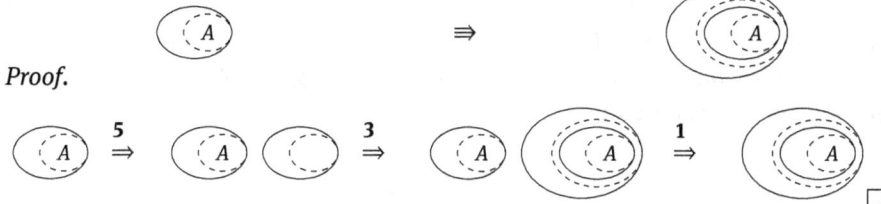

This deduction corresponds to the modal axiom $\Box A \rightarrow \Box\Box A$, which is characteristic of logic S4. With suitable rules of iteration and deiteration, we obtain systems of existential graphs for intuitionistic modal logics S4, S4.2, and S5.

3.2.3.2 Intermediate logics

An *intermediate logic*, also known as a *superintuitionistic logic*,[49] is a sublogic of classical propositional logic that contains IPL. Thus, an intermediate logic corresponds to adding a certain set of formulas to IPL. Since the set of formulas of IPL is denumerable, in principle there are as many intermediate logics as there are real numbers.

An intermediate logic is *finitely axiomatizable* if we can deduce all its theorems by adding a finite set of axioms to IPL. Classical propositional logic is finitely axiomatizable since, to obtain a complete axiomatization, it is enough to add one axiom to IPL, for example the law of double negation, or the law of excluded middle. In fact, since IPL has conjunction, any finitely axiomatizable intermediate logic has the form IPL + φ, which means that we add the formula φ to the axioms of IPL and consider the logic of all its theorems. Gödel showed an infinite

---

49 Chagrov and Zakharyaschev 1997, p. 109.

sequence of such formulas, each of which gives rise to a different logic.[50] Therefore, the family of finitely axiomatizable intermediate logics is countably infinite.

Building upon intuitionistic Alpha graphs we immediately obtain a system of existential graphs for each finitely axiomatizable intermediate logic. Because, if this logic has the form IPL + φ where φ is a formula in the language of IPL, let F be the intuitionistic Alpha graph that corresponds to φ. Then the intuitionistic Alpha graphs with their established rules, to which we add the graph F as an axiom that we may draw or erase in any area, is a system of existential graphs for IPL + φ. Thus an infinite family of different logics between classical propositional calculus and IPL now has suitable existential graphs.

Example 3.37. *Gödel–Dummett propositional logic*, also denoted LC, is the intermediate logic given by IPL + $(\alpha \to \beta) \vee (\beta \to \alpha)$. Hence, we obtain a system of existential graphs for this logic adding to the intuitionistic Alpha graphs the following diagrammatic axiom:

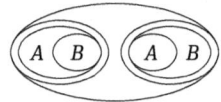

Example 3.38. $\underset{LC}{\Rightarrow} \neg A \vee \neg\neg A$.

Graphical statement:

Proof. From Example 3.30 we have $\neg\neg A \to \neg A \Rightarrow \neg A$, and from Example 3.31, applied to $\neg A$, we obtain $\neg A \to \neg\neg A \Rightarrow \neg\neg A$. Therefore:

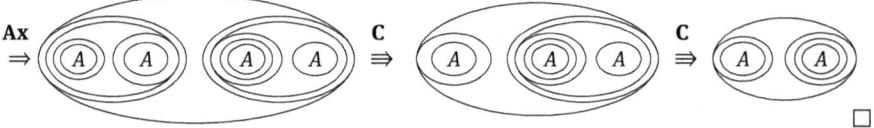

$\square$

In this way, we can express any finitely axiomatizable intermediate logic by means of existential graphs.

---

[50] Gödel 1932.

3.2.3.3 A lattice of intuitionistic subsystems

In another direction, from the intuitionistic Alpha graphs we also obtain systems of existential graphs for many sublogics of IPL. One way to get a sublogic is considering only some connectives and omitting the rest. In the graphical context, this means retaining some of the basic diagrams and eliminating others. Perhaps one of the simplest but interesting graphical systems emerges when we consider only the simple scroll on the sheet of assertion. Since this scroll represents the implication, we have this connective in the corresponding sublogic. But because juxtaposition on the sheet corresponds to conjunction, we naturally also have this connective. Thus we come to a prime example of a sublogic of IPL determined by connectives.

Implicational logic with conjunction (ILC) is the propositional logic whose only basic connectives are implication $\to$ and conjunction $\wedge$. The axioms of ILC are the same as those of IPL for these connectives:
1. $\alpha \to (\beta \to \alpha)$
2. $(\alpha \to (\beta \to \gamma)) \to ((\alpha \to \beta) \to (\alpha \to \gamma))$
3. $(\alpha \wedge \beta) \to \alpha$
4. $(\alpha \wedge \beta) \to \beta$
5. $(\gamma \to \alpha) \to ((\gamma \to \beta) \to (\gamma \to (\alpha \wedge \beta)))$

The only inference rule of ILC is *modus ponens* and the relation of entailment, which we now denote $\Sigma \vdash_{ILC} \varphi$, is defined exactly as in IPL. All properties referring only to implication and conjunction that are valid in IPL are also valid in ILC. Particularly, in this sublogic we have also the deduction theorem (Theorem 3.1). Corollary 3.1 acquires special relevance here, because it combines the only two connectives of ILC:

$$\gamma \vdash_{ILC} \alpha \to \beta \quad \text{if and only if} \quad \gamma \wedge \alpha \vdash_{ILC} \beta.$$

An adequate semantics for ILC is given by Hilbert semilattices. A *Hilbert algebra* is an algebraic structure $(H, \to, 1)$ that satisfies the following axioms:
1. $a \to (b \to a) = 1$
2. $(a \to (b \to c)) \to ((a \to b) \to (a \to c)) = 1$
3. If $a \to b = 1$ and $b \to a = 1$ then $a = b$

Every Heyting algebra is a Hilbert algebra. Any totally ordered set with maximum 1 is a Hilbert algebra, if we define:

$$a \to b = \begin{cases} 1 & \text{if } a \le b; \\ b & \text{otherwise.} \end{cases}$$

Hence, a totally ordered set with maximum but without minimum is a Hilbert algebra that is not a Heyting algebra.

In fact, any Hilbert algebra is partially ordered by the binary relation defined as:

$$a \le b \quad \text{if} \quad a \to b = 1,$$

and, for this order, the constant 1 becomes the maximum element. Like any partially ordered set, a Hilbert algebra may be or not a semilattice or even a lattice. A *Hilbert semilattice* is a Hilbert algebra that is a lower semilattice, that is, every pair of elements a, b has a greatest lower bound $a \wedge b$ with respect to the induced partial order. Furthermore, for all elements a, b, c, we require the following equivalence to hold:

$$c \le a \to b \quad \text{if and only if} \quad c \wedge a \le b,$$

which exactly matches Corollary 3.1 for ILC. We cannot deduce this additional condition from the other axioms of Hilbert algebras.

As in IPL, we may define valuations and the consequence relation in Hilbert semilattices, and then we obtain soundness and completeness theorems. In this way, these structures are the algebraic semantics of ILC. Thirdly, we can define a system of existential graphs for this logic.

The components from which the implicational Alpha graphs are built are:
- The plane surface, without border, upon which we draw all graphs, called the *sheet of assertion*;
- Propositions, symbolized by *capital letters*;
- Curves called *scrolls*, and composed of two simple closed curves, one of them inside the other and the two intersecting at only one point.

3 Intuitionistic and Geometrical Extensions of Peirce's Existential Graphs — 157

The outer curve of a scroll is called *cut* and the inner *loop*. The region limited by the cut and the loop is the *outer area* of the scroll and the interior of each loop is the *inner area*.

An *implicational Alpha graph* is a diagram composed of a finite combination of letters and scrolls, drawn upon the sheet of assertion. There may be repeated letters but they all occupy different places. The scrolls do not touch the letters nor do they touch each other. We consider two graphs that can be continuously deformed into each other as equal.

We derive the interpretation of the implicational Alpha graphs from the following clauses.
- The sheet of assertion is the universe of possibilities of truth.
- Drawing a graph on the sheet means asserting its interpretation. Writing a letter means asserting the proposition it represents.
- Drawing two graphs means asserting both.
- Drawing a scroll means asserting the implication whose antecedent is the graph in the outer area and whose consequent is the graph into the loop.

Consequently, these are the graphs for the basic connectives of ILC:

- Implication: $A \to B$ (A (B))

- Conjunction: $A \wedge B$ A B

Again, an *area* is a region of the sheet of assertion limited by curves, both cuts and loops. An area is *even* or *odd* if there is an even or odd number of curves around it, here we count both cuts and loops.

The following are the rules of transformation allowed for implicational Alpha graphs. Surprisingly, except the last one they have verbatim the same statements as in classical Alpha graphs.

1. *Erasure*. In an even area, any graph may be erased.

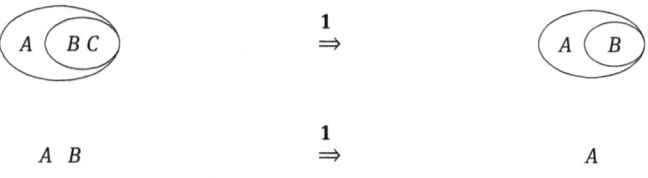

2. *Insertion.* In an odd area, any graph may be scribed.

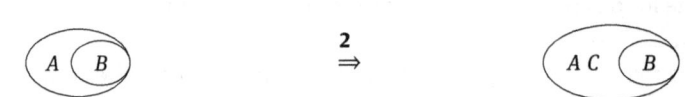

3. *Iteration.* Any graph may be iterated in its own area, or in any area contained in it, that is not part of the graph to be repeated.

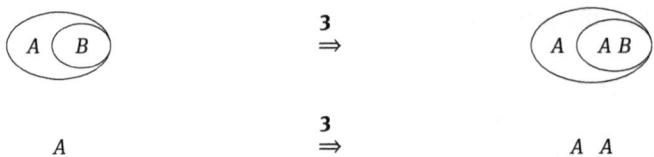

4. *Deiteration.* Any graph may be erased if a copy of it persists in the same area or in any area around it.

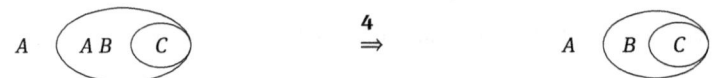

5. *Scrolling.* A scroll with empty outer area may be drawn around or removed from any graph on any area.

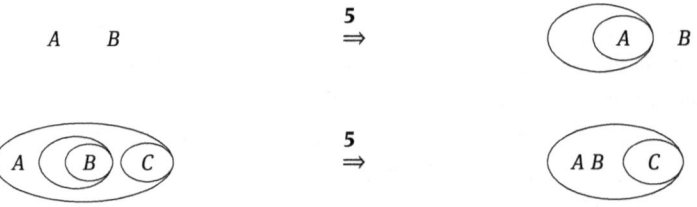

An implicational Alpha graph G entails graph H, and we write $G \underset{IC}{\Rightarrow} H$, if there exists a finite sequence of implicational Alpha graphs $A_1, A_2, \ldots, A_n$ (called a *deduction*) with $A_1 = G$ and $A_n = H$, such that every graph A_i is obtained from the preceding one A_{i-1} by the application of one of the five rules.

The same deductions given in III of *Subsection 3.2.2.2* apply to graphically prove *modus ponens*: $A, A \to B \underset{IC}{\Rightarrow} B$, and also $A \to B, B \to C \underset{IC}{\Rightarrow} A \to C$, $\underset{IC}{\Rightarrow} A \to (B \to A)$, and $\underset{IC}{\Rightarrow} A \to A$. We show one more example, which is also valid in intuitionistic graphs.

Example 3.39. $A \to (B \to C) \underset{IC}{\Longleftrightarrow} (A \wedge B) \to C$.

Graphical statement:

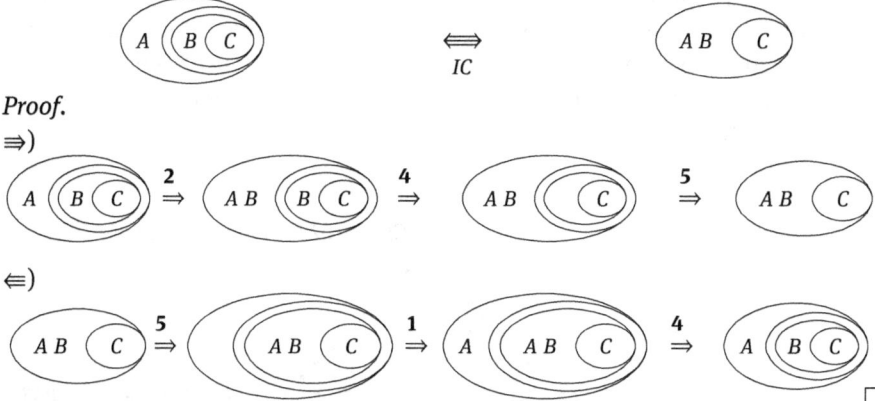

Proof.

$\Rightarrow$)

$\Leftarrow$)

$\square$

If again we add to this system the double scroll with its corresponding rules of transformation, we obtain a system of existential graphs for the *positive* sublogic of IPL, given by the connectives $\{\to, \wedge, \vee\}$. On the other hand, we could refine the graphical system for ILC admitting no juxtaposition of graphs within the areas of any scroll. Adjusting the rules of transformation accordingly, we obtain a system of existential graphs for *pure implicational logic*, the sublogic of IPL determined by the single connective $\{\to\}$, whose algebraic semantics is given by the Hilbert algebras mentioned above.

To any of these three sublogics we can add the negation in two steps. First, we consider a constant but without any additional axioms. In sublogics given by formulas, the constant is $\bot$ and we define $\neg \alpha$ as $\alpha \to \bot$, while in graphical systems the constant is the empty cut and we define the cut around an arbitrary graph G as the scroll with G in its outer area and the empty cut in the inner. Even without the principle of explosion, in this sublogics we can prove –both algebraically and graphically– several interesting properties of negation, such as *modus tollendo tollens*.

For the second step, we add the principle of explosion. In the formulas, we achieve this by adopting the axiom $\bot \to \alpha$, and in the graphs, we extend the insertion of loops to the empty cut and, therefore, to all simple cuts. Thus, for any of the three implicational sublogics determined by $\{\to\}$, $\{\to, \wedge\}$, and $\{\to, \wedge, \vee\}$, we also obtain three sublogics, namely, the positive logic, the sublogic with the constant and weak negation, and the system with the complete negation. One of these nine sublogics is the complete IPL, and all nine have an associated system of existential graphs.

In the following diagram, each dot • is a propositional logic for which there is a system of existential graphs similar to Peirce's classical Alpha graphs. The top diamond loosely represents the countably infinite logics between classical and intuitionistic propositional logics that are finitely axiomatizable. The bottom diamond displays the nine sublogics of IPL discussed above, each dot is labeled with the connectives that determine the corresponding logic. Here ⌐ denotes the constant ⊥ without the principle of explosion, which in the graphs corresponds to the empty cut without the insertion of loops.

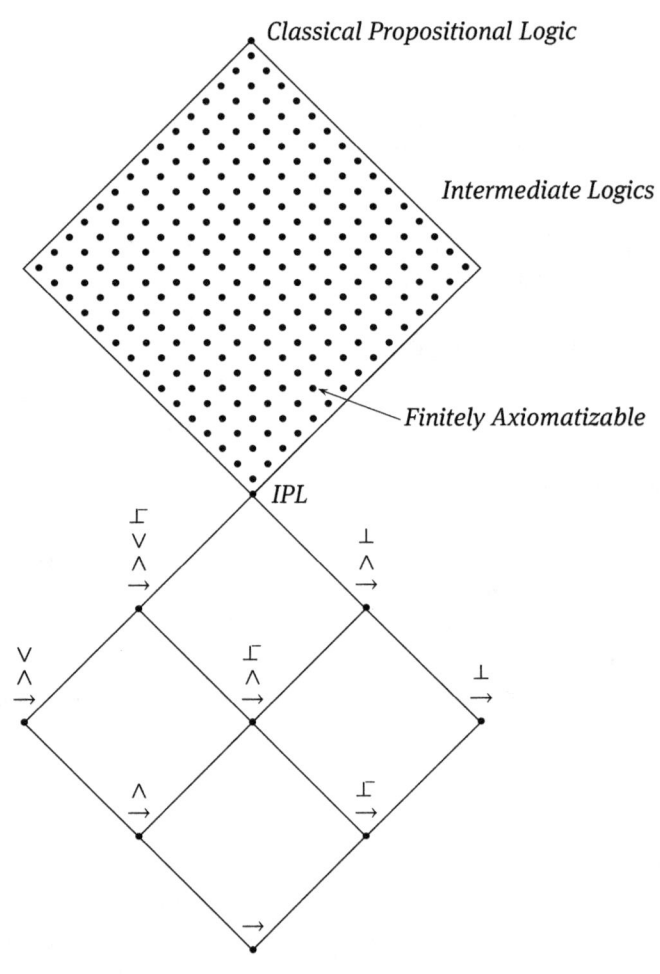

This map clearly points to intuitionistic logic as the center of the current realm of logics with which a matching system of existential graphs is known to be associated. Classical logic becomes an extreme case in this field, the other one being pure implicational logic.

Adding *Peirce's law* $((\alpha \to \beta) \to \alpha) \to \alpha$ to pure implicational logic, we obtain *classical implicational propositional logic*. Proceeding as in the case of finitely axiomatizable intermediate logics, by adding the graphical version of this law as an axiom to the graphs for pure implicational logic, we obtain a system of existential graphs for classical implicational logic. In this logic we have neither conjunction nor negation but, again, adding the negation in two steps we recover the classical propositional logic. Therefore, for these two sublogics of classical propositional logic, classical implicational logic and classical implicational logic with constant, there are also systems of existential graphs.

A still open problem in the study of systems of existential graphs for intuitionistic sublogics consists in examining logics with disjunction but that do not have both of the other binary connectives. This would lead us to three other sublogics, and with negation we would obtain another lattice of nine sublogics. Finally, the logics with only conjunction or with no connectives at all seem to have fairly trivial systems of existential graphs. Again with negation, we might add other six systems. Thus, there is a grand total of twenty-four sublogics of IPL that could have a diagrammatic version in the form of a system of existential graphs similar to Peirce's Alpha graphs.

To any of the logics mentioned in this section for having a system of Alpha graphs fitting with its propositional part, we can also associate Beta graphs and the various modal Gamma graphs systems. Thus, existential graphs are permeating the realm of non-classical logic at many levels.

Postscript

Although they are a very small minority in the mathematical world, there have been some people, mostly disciples of Brouwer, who have advanced pure intuitionism.[51] A significant interest in Brouwer's life and thought also persists.[52] On the other hand, intuitionistic logic plays an important role in the field of non-classical logics,[53] while Heyting algebras appear recurrently in lattice theory.[54]

---

[51] See, for example, Heyting 1971, and Troelstra and van Dalen 1988.
[52] van Dalen 2013, van Stigt 1990, and Largeault 1993.
[53] See, *e.g.*, Priest 2008, Chagrov and Zakharyaschev 1997, or Goldblatt 1979.
[54] Oostra 1997, Blyth 2005, and L. Acosta 2015.

The possibility of developing intuitionistic existential graphs was first suggested by Zalamea in the 1990s.[55] Existential graphs are a true geometric and topological expression of logic, and Tarski pointed out the close relation between topology and intuitionistic logic.[56] Starting in 2007, Oostra proposed the system of intuitionistic Alpha graphs presented in this section, and published it in 2010.[57] He also developed the natural extension to Beta and Gamma intuitionistic existential graphs.[58] With his students, he inquired into the sublogics of IPL[59] and the intermediate logics,[60] and also developed the graphical system for implicational logic with conjunction[61] and pure implicational logic.[62] In addition, they developed a proof method for the equivalence between systems of Alpha graphs and the syntactic presentations of propositional logics.[63] This finally led to the formal equivalence proof for intuitionistic Alpha graphs.[64]

3.3 Existential Graphs on Surfaces

Existential graphs constitute an authentic geometrization of logic and, through them, logic can benefit from the versatility of geometry. In Peirce's own words, one of the basic ideas of his graphs consists in "spreading the formulae over two dimensions".[65] Geometry has shown that two-dimensionality is not restricted to the plane, and we can draw existential graphs very naturally on nonplanar surfaces such as the sphere, the cylinder, or the torus. In the same way as the results of geometry change on different surfaces, the systems expressed by existential graphs on nonplanar surfaces do not obey the classical rules.

Thus a surprising new interplay between logic and geometry is born. In one direction, given a logic, we may ask whether there is an adequate system of existential graphs for it. As we have seen, in many cases the answer is yes. Now, in

---

55 Zalamea 1997a, and Zalamea 1997b.
56 Tarski 1938.
57 Oostra 2010.
58 See Oostra 2011, and Oostra 2012.
59 Castillo 2009, and Castillo and Oostra 2010.
60 Moreno 2014.
61 Gómez 2013, see also Oostra 2019b.
62 F. Calderón and M. Calderón 2021.
63 Gómez 2013, and Fuentes 2014.
64 Ortiz and Segura 2018, and Oostra 2021.
65 [1982–2009] W 4.394.

the other direction, given a surface, we may ask if there is a logical system that corresponds to the existential graphs drawn on it.

3.3.1 Alpha graphs on surfaces

In this subsection, we define existential Alpha graphs on any two-dimensional surface. Again, here we assume the basic notions of general topology.

3.3.1.1 Surfaces

In an informal way, a surface is a mathematical object that *locally* —that is, in any of its "small" parts— is like a plane, but which does not necessarily extend unlimitedly flat and level in all directions like the plane, but that *globally* can take many different shapes.

With more technical precision, a surface is a connected topological 2-manifold. This is a connected Hausdorff topological space with an open cover, each of whose members is homeomorphic to some open subset of the Euclidean plane, that is, the plane with its usual topology. Such a homeomorphism is known as a *chart*, hence the cover is an *atlas*. With symbols, a surface S is a connected Hausdorff space plus an atlas $\{\varphi_i\}_i$, here for each i the chart $\varphi_i : U_i \to V_i$ is a homeomorphism, U_i is an open subset of S, V_i is an open subset of $\mathbb{R}^2$, and finally $\bigcup_i U_i = S$.

Example 3.40. We can define the *sphere*, denoted S^2, as the set of all points of Euclidean space that are at a distance 1 from the origin, that is, all $(x, y, z) \in \mathbb{R}^3$ that satisfy

$$x^2 + y^2 + z^2 = 1,$$

with the subspace topology. The sphere is also described by the parametric equations:

$$x = \cos \lambda \cos \theta, \qquad y = \cos \lambda \sin \theta, \qquad z = \sin \lambda,$$

here λ is the *latitude*, with $-\frac{\pi}{2} \leq \lambda \leq \frac{\pi}{2}$, and θ is the *longitude*, with $-\pi \leq \theta \leq \pi$. These equations define not an injective mapping, but its restriction $\alpha_1 : (-\frac{\pi}{2}, \frac{\pi}{2}) \times (-\pi, \pi) \to S^2$, given by $\alpha_1(\lambda, \theta) = (\cos \lambda \cos \theta, \cos \lambda \sin \theta, \sin \lambda)$, is a homeomorphism from an open set of the plane onto an open subset of the sphere. Taking $\alpha_2 : (-\frac{\pi}{2}, \frac{\pi}{2}) \times (-\pi, \pi) \to S^2$ defined as $\alpha_2(\lambda, \theta) = (-\cos \lambda \cos \theta, \sin \lambda, \cos \lambda \sin \theta)$ the two images cover the sphere, hence their inverses $\{\alpha_1^{-1}, \alpha_2^{-1}\}$ constitute an atlas for the sphere.

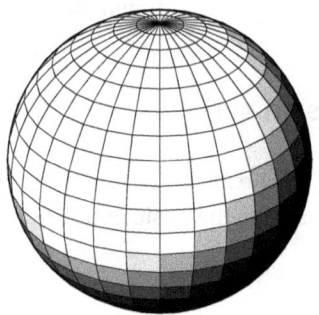

The well-known *stereographic* projection of the sphere consists in mapping any point different than the *north pole* $n = (0, 0, 1)$ on the plane along the line that passes trough n and the point, precisely, $\varphi_n : S^2 \to \mathbb{R}^2$ is given by $\varphi_n(x, y, z) = \left(\frac{x}{1-z}, \frac{y}{1-z}\right) = \frac{1}{1-z}(x, y)$, with inverse $\alpha_n(x, y) = \varphi_n^{-1}(x, y) = \left(\frac{2x}{x^2+y^2+1}, \frac{2y}{x^2+y^2+1}, \frac{x^2+y^2-1}{x^2+y^2+1}\right)$. This φ_n is a homeomorphism of the open set $S^2 \setminus \{n\}$ onto the whole plane, which is an open set. The symmetric projection from the *south pole* $s = (0, 0, -1)$ given by $\varphi_s(x, y, z) = \left(\frac{x}{z+1}, \frac{y}{z+1}\right) = \frac{1}{z+1}(x, y)$, whose inverse is given by $\alpha_s(x, y) = \left(\frac{2x}{1+x^2+y^2}, \frac{2y}{1+x^2+y^2}, \frac{1-x^2-y^2}{1+x^2+y^2}\right)$, completes another atlas $\{\varphi_n, \varphi_s\}$ for the sphere.

Since the sphere is compact and the homeomorphic image of a compact set is compact, and because there is no non-empty compact open set in the plane, there is no atlas with less than two charts for the sphere.

Example 3.41. The *cylinder* is the pipe that extends indefinitely in both directions. We can describe it as the set of all points (x, y, z) in Euclidean space that satisfy

$$x^2 + y^2 = 1,$$

with the subspace topology, or by the parametric equations:

$$x = \cos\theta, \qquad y = \sin\theta, \qquad z = \lambda,$$

where $\lambda \in \mathbb{R}$ takes any real value, while $-\pi \leq \theta \leq \pi$. The mapping defined as $\varphi(x, y, z) = (xe^z, ye^z) = e^z(x, y)$ is a homeomorphism from the cylinder to the open subset $\mathbb{R}^2 \setminus \{0\}$, with inverse $\alpha(x, y) = \left(\frac{x}{\sqrt{x^2+y^2}}, \frac{y}{\sqrt{x^2+y^2}}, \ln\sqrt{x^2+y^2}\right)$. In this case, $\{\varphi\}$ is an atlas with a single chart.

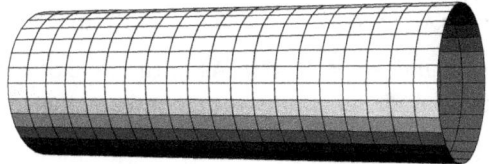

We consider here the cylinder without border, hence it is equivalent to the bounded cylinder with, for example, $-\frac{1}{2} < \lambda < \frac{1}{2}$.

Example 3.42. The *Möbius strip* is the result of joining the ends of a rectangle, giving one of them a half-twist. The Möbius strip without borders, of width 1 and around the unit circle, has parametric equations:

$$x = \left(1 + \lambda \cos \frac{\theta}{2}\right) \cos \theta \qquad y = \left(1 + \lambda \cos \frac{\theta}{2}\right) \sin \theta \qquad z = \lambda \sin \frac{\theta}{2},$$

where $-\frac{1}{2} < \lambda < \frac{1}{2}$ and $-\pi \leq \theta \leq \pi$. This is the simplest non-orientable surface, in the sense that it not possible to define on it a consistent notion of "clockwise" rotation.

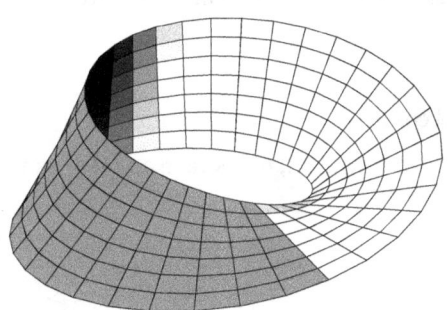

Example 3.43. The *torus* is a surface of revolution generated by rotating a circle about an axis that is coplanar with the circle and does not touch it. We can describe it as the set of all points (x, y, z) in Euclidean space that satisfy

$$\left(\sqrt{x^2 + y^2} - R\right)^2 + z^2 = r^2,$$

with the subspace topology. Here r is the radius of the rotating circle, R is the radius of revolution of its center, and we assume $0 < r < R$. The torus is also given by the parametric equations:

$$x = (R + r\cos\lambda)\cos\theta \qquad y = (R + r\cos\lambda)\sin\theta \qquad z = r\sin\lambda,$$

where $-\pi \leq \lambda \leq \pi$ and $-\pi \leq \theta \leq \pi$. As in the case of the sphere, with some adjustments to this parameterization we can find an atlas for the torus.

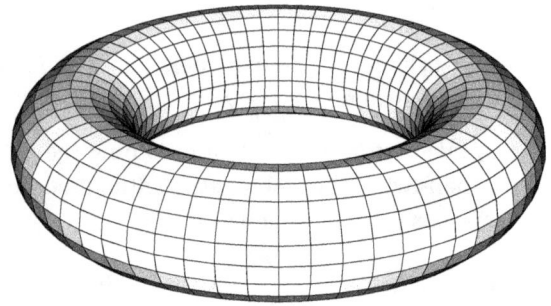

Example 3.44. The *helicoid* is a ruled surface described by a line that rotates at constant rate around an axis that is perpendicular to it, and at the same time moves at a constant rate along this axis. The parametric equations for the helicoid are:

$$x = \lambda\cos\theta \qquad y = \lambda\sin\theta \qquad z = k\theta,$$

where $k \neq 0$ is constant, while $\lambda, \theta \in \mathbb{R}$ take any real value.

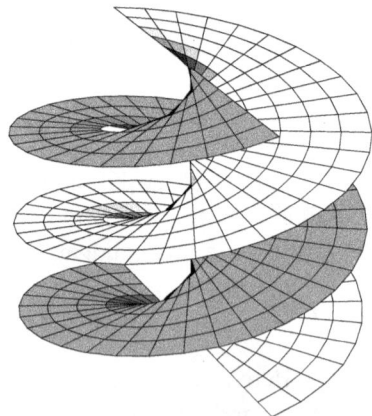

The variant helicoid with $\lambda > 0$ and $k = 1$ is homeomorphic to the whole plane through the complex exponential function.

There are some interesting classification theorems for surfaces. A surface is *closed* if it is compact and without boundary; in our previous examples, only the sphere and the torus are closed in this sense. Any closed surface is homeomorphic to exactly one member of these three families:
1. the sphere,
2. the connected sum of g tori for $g \geq 1$, this number g is the *genus* of the surface,
3. the connected sum of k real projective planes for $k \geq 1$.

The surfaces of the first two families are orientable while those of the third are non-orientable.

3.3.1.2 Definition of Alpha graphs

To consider Alpha graphs on an arbitrary surface, it is actually enough to interchange the flat sheet of assertion for the chosen surface. Hence, the components from which the Alpha graphs on a surface S are built are:
- The *surface S* upon which we draw all graphs, which now plays the role of the sheet of assertion;
- Propositions, symbolized by *capital letters*;
- Simple closed curves, called *cuts*.

An *Alpha graph* on S is a diagram composed of a finite combination of letters and cuts, drawn upon the surface S. There may be repeated letters but they all occupy different places. The cuts do not touch the letters nor do they touch each other. We consider two graphs that can be continuously deformed into each other as equal.

As in the original case on the plane, we can upgrade this descriptive definition of Alpha graphs on the surface S to a formal inductive definition, although on particular surfaces various feasible forms of such definitions may arise. On the other hand, we can always work out a precise definition of these graphs as mathematical objects, just as we did in the first section of this chapter. An Alpha *pre-graph* on S is a continuous and injective map $\alpha : mS^1 + F_n \to S$ with an arbitrary labeling function $\lambda : F_n \to L$. Again, m and n are non-negative integers, mS^1 is the topological sum of m copies of the unit circle, $F_n = \{1, 2, \ldots, n\}$ is a finite set with the discrete topology, and L is the set of propositional letters considered. Under these circumstances, α is an embedding, the letters occupy different places, and the cuts do not touch the letters or touch each other.

Two pre-alpha graphs ($\alpha : mS^1 + F_n \to S, \lambda$), ($\alpha' : m'S^1 + F_{n'} \to S, \lambda'$) are *equivalent* if $m = m'$ and $n = n'$ —therefore, we can assume that the domains are the same—, also $\lambda = \lambda'$, and, in addition, there exists an isotopy $Y : \alpha \to \alpha'$. This is an equivalence relation in the set of pre-graphs, and we define an *Alpha graph* on S as an equivalence class, or isotopy class, of Alpha pre-graphs. In this way, every Alpha graph on the surface is a well-defined mathematical object.

In any system of existential graphs, we can distinguish three fundamental features: the graphs in themselves—the *syntax*—, their logical interpretation—the *semantics*—, and the rules of transformation—the *pragmatics*—. These three together lead to a deduction relation, which expresses a certain logic. From the above account, we can conclude that Alpha graphs have an accurate description on any topological surface. Therefore, the open problem on any particular manifold lies in the interpretation given to these graphs and the adoption of appropriate rules of transformation. These features depend, in turn, essentially on the geometric properties of the Jordan curves on the surface, and must be invariant under isotopies of pre-graphs. In this way, the geometry of the chosen manifold may—or may not—determine a certain logic on it by means of existential graphs.

3.3.2 Some case studies

For an interpretation of Alpha graphs on any surface, we hold to the general notion that the sheet of assertion is the universe of possibilities of truth. A whole graph, drawn on the surface, is thus asserted. Furthermore, writing a letter on the surface means asserting the proposition it represents, and writing two letters means asserting both of them. Hence, the main issue in the interpretation of Alpha graphs on nonplanar surfaces lies in the meaning given to the cut.

In this subsection, we will ponder Jordan curves on different surfaces and propose possible interpretations of Alpha graphs on them.

3.3.2.1 The sphere

Geometrically, a Jordan curve on the sphere divides it into two components that are both bounded, moreover, they are topologically indistinguishable. Therefore, a cut on the sphere does not *enclose* an area, a letter, or a graph, and consequently we cannot use it to express something about the enclosed graph, such as its falsity or non-contingency, as we do on the plane. On this surface, a cut only *separates* the sheet of assertion into two areas, thence we can use it to express something about the two graphs set apart. Thus, in any interpretation of the cut, we switch from asserting a monadic predicate to expressing a binary relation.

In principle, just as on the plane, there are several possible or imaginable interpretations for the cut on the sphere. In the very basic context of an *opposition* relation,[66] we may define that two letters on the sphere *disagree* if they are separated by a cut drawn on the surface, and they *agree* otherwise. Or, in an interpretation of *discernability* or *discrimination*, we can define that two letters on the sphere are *distinguishable* or *distinct* if they are separated by a cut and that they are *identical* or *similar* in the opposite case.

Example 3.45. For an opposition relation, this graph on the sphere asserts that letters A and B disagree, while C and D agree.

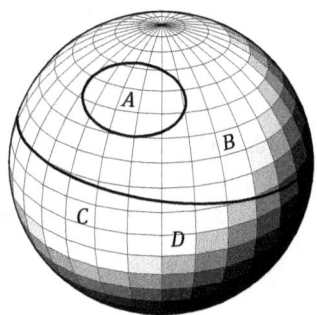

This example raises the question of what happens when a graph on the sphere includes various cuts. If, as in this case, A disagrees with B, and B disagrees with C, what about A and C? Actually, there is a technical basis for the somewhat classical idea of identifying a *double disagreement* with an agreement, because the relation of crossing an even number of cuts in passing from one point to another is well-defined on the sphere, and divides it into *two* regions. Some details follow.

Let us consider a finite family $C_1, C_2, \ldots, C_n$ of disjoint Jordan curves on the sphere, with $n \geq 1$. For any pair of points x, y in its complement $S^2 \setminus (C_1 \cup \cdots \cup C_n)$, we define n numbers $\sigma_i(x, y)$ as follows:

---

[66] Referring to *simple relatives*, Peirce wrote: "The former express a mere agreement among things, the latter set one thing over against another, and in that sense express an opposition (ἀντικεῖσθαι); I shall therefore term the former *concurrents*, and the latter *opponents*." [1982–2009] W 2.418.

$$\sigma_i(x, y) = \begin{cases} 1 & \text{if } x, y \text{ belong to different components of the curve } C_i, \\ 0 & \text{otherwise.} \end{cases}$$

Next, we define the binary relation ~ on $S^2 \setminus (C_1 \cup \cdots \cup C_n)$ as:

$$x \sim y \quad \text{if} \quad \sum_{i=1}^{n} \sigma_i(x, y) \text{ is even.}$$

This is an equivalence relation with exactly two equivalence classes.[67] The same result holds on the plane—where it indicates the odd and even areas of an Alpha graph—and the cylinder, but not on all surfaces. For example, it fails on the torus and the Möbius strip.

This relation $x \sim y$ does not change for points x', y' that belong to the same component as x, y, respectively, therefore we can define the relation $X \sim Y$ for connected components X, Y of $S^2 \setminus (C_1 \cup \cdots \cup C_n)$. In the context of an Alpha graph on the sphere, two areas X, Y are *evenly separated* if $X \sim Y$ for the curves of the graph, and *oddly separated* otherwise. Also, for letters A, B of such a graph, $A \sim B$ means that they are written in evenly separated areas. Thus, in the following table, X and Y may be areas or letters, while $\not\sim$ denotes the (metalogical) negation of the relation ~.

$X \sim Y$	$Y \sim Z$	$X \sim Z$
$X \sim Y$	$Y \not\sim Z$	$X \not\sim Z$
$X \not\sim Y$	$Y \sim Z$	$X \not\sim Z$
$X \not\sim Y$	$Y \not\sim Z$	$X \sim Z$

This table has a strong, although merely formal, resemblance to the truth table of the biconditional connective of classical propositional logic. If we write $x \Delta y$ for $x \not\sim y$ and the same for areas and letters, then for this relation Δ we obtain the following table.

$X \Delta Y$	$Y \Delta Z$	$X \Delta Z$
$X \Delta Y$	$Y \blacktriangle Z$	$X \Delta Z$
$X \blacktriangle Y$	$Y \Delta Z$	$X \Delta Z$
$X \blacktriangle Y$	$Y \blacktriangle Z$	$X \blacktriangle Z$

---

67 For details, see Niño 2021.

This second table, in turn, strongly resembles the exclusive disjunction truth table.

Thus, the following basic clauses give at least one feasible interpretation of the Alpha graphs on the sphere.
- Drawing a graph on the sphere means asserting its interpretation.
- Writing a letter on the sphere means asserting the proposition it represents.
- Writing two letters on evenly separated areas of the sphere means asserting they agree, and writing them on oddly separated areas means they disagree.

This proposed interpretation does not automatically extend to graphs. In fact, the notion of subgraph is not easily defined in the context of the sphere.

Even more than the interpretation of the graphs on the sphere, the formulation of suitable rules of transformation constitutes a real drawback in investigations of existential graphs on this surface. Since a cut divides the sphere into two identical hemispheres, we cannot naturally and consistently point to either area as the outside or the inside. Hence, there is no inward direction and we cannot transfer the rules of iteration and deiteration to the sphere. Furthermore, although different cuts divide the sphere into two equivalence classes, there is none that we can choose naturally and consistently as the outside, and thus define odd and even areas. Hence, we cannot directly adapt the rules of erasure and insertion to the sphere.

These are the only rules of transformation that we may convey directly to the sphere:
3. *Iteration.* Any letter may be iterated in its own area.
4. *Deiteration.* Any letter may be erased if a copy of it persists in the same area.
5. *Double cut.* A double cut, made up of two cuts that delimit a cylindrical area without letters or cuts, may be drawn on or erased from any area.

With these rules only, it seems difficult to attain logically significant proofs. Future efforts should focus on entirely new rules of transformation that allow arbitrary letters to be erased or inserted in an Alpha graph on the sphere. Another point of investigation is the extension of these rules to arbitrary graphs, for which a precise notion of subgraph is required.

There is a completely different approach to Alpha graphs on the sphere that does yield a deduction relation. By choosing any point on the sphere, its complement is homeomorphic to the plane through a rotation and the stereographic projection. Hence, given an Alpha graph on the sphere, any point on its complement gives a *planar reading* of the graph as an usual Alpha graph on the plane sheet of assertion. Since this reading does not depend on the chosen point in a given area, any Alpha graph on the sphere has a finite number of planar readings.[68] Now, an Alpha graph G on the sphere *entails* graph H if each planar reading of G entails the corresponding planar reading of H, according to the usual rules of transformation.

Example 3.46. This is a spherical version of *modus ponens*:

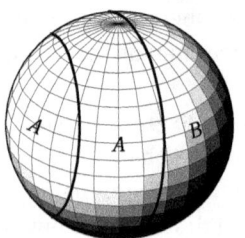 entails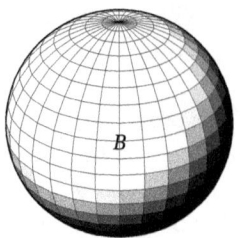

The three planar readings of the first graph are:

$$\begin{cases} A\;(A\;B) \stackrel{4}{\Rightarrow} A\;(B) \stackrel{5}{\Rightarrow} A\;B \stackrel{1}{\Rightarrow} B \\ (A\;(A\;B)) \stackrel{4}{\Rightarrow} (\;(A\;B)) \stackrel{1}{\Rightarrow} (\;) \stackrel{2}{\Rightarrow} (B) \stackrel{5}{\Rightarrow} B \\ (A)\;A\;B \stackrel{1}{\Rightarrow} B \end{cases}$$

Perhaps this local geometric entailment may shed light on the global rules of transformation for Alpha graphs on the sphere.

3.3.2.2 The cylinder

A Jordan curve on the cylinder divides it into two components, but two different classes of these curves arise on this surface. In one case, such a curve divides the cylinder into a bounded and an unbounded component, as on the plane. But other Jordan curves surround the cylinder and divide it into two components, both unbounded and topologically indistinguishable, as on the sphere. We will call one of the first type a *contractible* curve, because the map from the unit circle to

---

68 For more details, see Niño 2021.

the cylinder that draws it is homotopic to a constant map from the circle to the cylinder, while in the second case there is no such homotopy, so the curve is *non-contractible*.

On the cylinder, a contractible cut encloses an area and we might use it to assert a monadic predicate, such as negation, while a non-contractible cut only separates two areas of the sheet of assertion and we might express with it a binary relation, as opposition.

Example 3.47. This graph on the cylinder asserts that letters A and B disagree, while it negates proposition C.

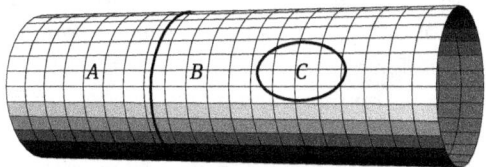

As on the plane and the sphere, any finite family of disjoint Jordan curves on the cylinder defines an equivalence relation in its complement, with exactly two classes. Moreover, counting both kinds of cuts equally, every area is odd or even with respect to each of the unbounded ends of the cylinder. This parity is the same with respect to the two extremes if there is an even number of non-contractible cuts on the cylinder, but exactly the opposite in the other case. This could give us a somewhat classical interpretation of the Alpha graphs on the cylinder, with truth values of all propositions with respect to two "absolute" truths that may or may not be consistent. Again, the only suitable rules of transformation for this interpretation are iteration and deiteration in the same area, and double cut.

The graphs on the cylinder are greatly simplified when there are no non-contractible curves, that is, when we restrict ourselves to Alpha graphs with contractible cuts. In this situation, each cut has an inside and an outside, and the exteriors of different cuts always have points in common, so there is only one outside for the whole graph, and the inside direction is consistent for different cuts. Therefore, we can read the graph with the same interpretation of Alpha graphs on the plane. Moreover, we can consistently define the parity of the areas as on the plane, and we may apply all the usual rules of transformation, even enclosing any of these graphs in a double cut made up of non-contractible cuts. Thus we obtain the following result.

Theorem 3.6. The system of *contractible Alpha graphs on the cylinder*, composed of the Alpha graphs whose cuts are all contractible, with the same interpretation and rules of transformation as on the plane, is a graphical version of classical propositional logic.

In the same way, we might consider Beta graphs, modal Gamma graphs, and even intuitionistic existential graphs on the cylinder, exactly as on the plane. The only restriction is the absence of non-contractible cuts.

In the general case, non-contractible cuts divide the cylinder into a finite number of sub-cylinders. The content of each of these cylinders is a contractible Alpha graph and, by Theorem 3.6, we can interpret it as an Alpha graph on the plane. In this way, we may consider an Alpha graph on the cylinder as a finite set of classical Alpha graphs separated by non-contractible cuts. We might interpret these cuts as opposition, or perhaps only as adjacency, or even as conjunction, of the different graphs. The interpretation chosen would lead us to suitable rules of transformation for the Alpha graphs on the cylinder.

An altogether different use of non-contractible cuts on the cylinder is Peirce's decision method for Alpha graphs.[69] Given any Alpha graph G on the plane, we now start the procedure with a cylinder, separated in two by a non-contractible cut and containing the graph G in one of its components. The only non-contractible cut plays the role of the dotted line introduced in *Subsection 3.1.1.2*. On this graph, we perform the five operations described in that part, taking as the "lower region" the component of the cylinder where G was originally drawn, and the other as the "upper region". In operation 5 we take a copy of the entire cylinder. In the end, the disjunction of the Alpha graphs left on the upper regions is equivalent to the original graph G. From this, we can determine the combinations of letters that make G true, or we may obtain the disjunctive normal form of G.

Finally, we could consider *multiple cylinders*. By stereographic projection, the plane is homeomorphic to the sphere minus one point, also known as the *punctured sphere*. In the same way, the cylinder is homeomorphic to the twice-punctured sphere. Going further, we could consider the sphere minus any finite number of points, which is homeomorphic to the same number of half-cylinders diverging in different directions, all joined at the center. On such a surface, Theorem 3.6 still holds. However, in a general interpretation, any area of a graph has a truth value with respect to each of the unbounded ends of the half-cylinders, and the non-contractible cuts describe the coherence between the different "absolute" truths.

---

[69] This idea was proposed first on the torus, *cfr.* Arana 2020, but it works still better on the cylinder.

3.3.2.3 The torus

Two classes of Jordan curves arise on the torus, which we can describe again as contractible and non-contractible. A contractible Jordan curve divides the torus into two components, that are both bounded because the whole surface is bounded. A non-contractible Jordan curve *does not divide* the torus, and its complement is homeomorphic to the cylinder. For this second option, there are infinite different choices: through the hole of the torus, or along the torus and around its hole, or turning once along it while going through the hole many times, and so on. However, in all these cases the torus is not divided and the complement of the curve is homeomorphic to a cylinder. Consequently, a given finite family of disjoint Jordan curves may not define a separation equivalence relation. All we can say is that for such a family all non-contractible curves are of the same type and that they divide the torus into an equal number of cylinders.

Although both components given by a contractible Jordan curve on the torus are bounded, they are topologically different since one has a hole and the other does not. Components with a hole of two contractible curves always have points in common, hence we may define this as the outside of the curve while the other component is its inside. Thus, like the contractible Jordan curves on the cylinder and like all such curves on the plane, the contractible Jordan curves on the torus enclose a well-defined region of the surface. Therefore, we can use a contractible cut on the torus to express the negation of its interior, while a non-contractible cut seems to lack any possible semantical interpretation.

Example 3.48. This graph on the torus denies proposition A but apparently does not express any relation between letters B and C.

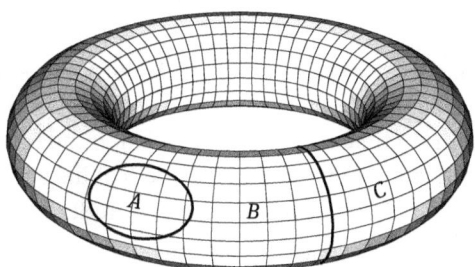

If we restrict ourselves to Alpha graphs on the torus with contractible cuts, the inside direction is well-defined, we can consistently define area parity as on the plane, and thus we may apply all the usual rules of transformation.

Theorem 3.7. The system of *contractible Alpha graphs on the torus*, composed of the Alpha graphs whose cuts are all contractible, with the same interpretation and rules of transformation as on the plane, is a graphical version of classical propositional logic.

Again, we might consider Beta graphs, modal Gamma graphs, and even intuitionistic existential graphs on the torus.

In the general case, the non-contractible cuts of an Alpha graph divide the torus into a finite number of cylinders. The content of each of these portions is a contractible Alpha graph and, by Theorem 3.6, we can interpret it as an Alpha graph on the plane. Hence, as on the cylinder, on the torus we may consider an Alpha graph—if it has more than one non-contractible cut—as a finite set of classical Alpha graphs separated by non-contractible cuts. In this case, we might interpret these cuts as conjunction, or merely as adjacency, of the classical graphs, but perhaps not as opposition because a parity relation seems not to be definable on the torus. In which cases might a first non-contractible cut be added? The interpretation and the chosen rules would lead us to a full understanding of the Alpha graphs on the torus.

Looking forward to subsequent investigations, in the future we may consider Alpha graphs on the connected sum of various tori, and also on a punctured torus.

3.3.2.4 The Möbius strip

On the Möbius strip, there are also contractible and non-contractible Jordan curves. A contractible Jordan curve on this surface divides it into two components, one bounded and the other unbounded, as occurs on the plane and on the cylinder. But here two different kinds of non-contractible Jordan curves emerge. A curve parallel to the central circle of the Möbius strip and at a constant distance from this circle will make two turns along the Möbius strip—and around its hole—and divides the surface into two regions. One of them is a bounded Möbius strip containing the central circle, the other is an unbounded cylinder. The central circle, however, only makes one turn along the Möbius strip and *does not divide* it, while its complement is homeomorphic to a cylinder.

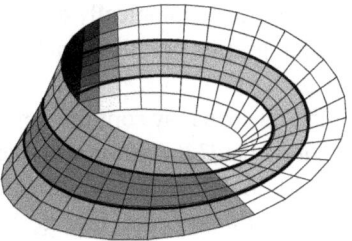 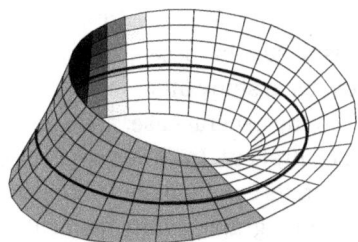

We will call *central* a Jordan curve on the Möbius strip that is homotopic to the central circle. A finite family of disjoint Jordan curves on this surface that contains a central curve does not define a separation equivalence relation. On the other hand, any non-central Jordan curve on the Möbius strip—even if it is non-contractible—has a bounded inside and an unbounded outside, thus enclosing a well-defined region of the surface. Hence, we can use any non-central cut on the Möbius strip to express the negation of its interior.

Example 3.49. The following Alpha graphs on the Möbius strip both deny A, but the one on the right apparently does not express any relation between letters B and C.

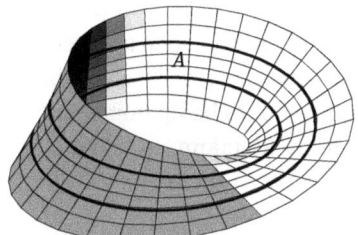 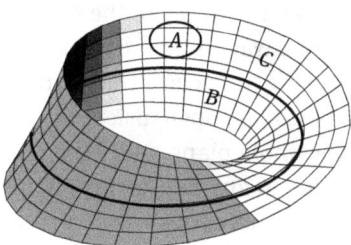

For an Alpha graph on the Möbius strip without any central cut, there is only one outside for the whole graph, each cut has a well-defined inside and outside, and the inside direction is consistent for different cuts. Hence, we can read it with the usual interpretation of Alpha graphs on the plane, we can consistently define the parity of the areas, and we may apply all the usual rules of transformation to it, even enclosing in a double cut made up of cuts that are not central.

Theorem 3.8. The system of *noncentral Alpha graphs on the Möbius strip*, composed of the Alpha graphs without any central cut, with the same interpretation and rules of transformation as on the plane, is a graphical version of classical propositional logic.

With this restriction, we might also consider all other known systems of existential graphs on the Möbius strip. Moreover, since there are two topologically different kinds of cuts in these graphs, perhaps the existential graphs on this surface will allow us to study other logics.

In the general case, an Alpha graph on the Möbius strip may contain at most one central cut. In which cases might it be added or deleted? If a graph has a central cut, its complement is homeomorphic to the cylinder, hence we can consider the Alpha graph minus the central cut as a graph on the cylinder, that is, as a finite set of classical Alpha graphs separated by non-contractible cuts. The interpretation and the chosen rules would lead us to a full understanding of all Alpha graphs on the Möbius strip.

In the future, we might also consider Alpha graphs on the Klein bottle, on the real projective plane, and on the connected sum of various projective planes.

3.3.3 The way ahead

In this final part, we clear out some ideas that arose in the case studies of specific surfaces and that may prove useful for general considerations about systems of Alpha graphs on surfaces, and we pose some open problems.

The complement of a chosen point on the sphere is homeomorphic to the plane, and also any of the two components of a chosen cut. We can interpret in the usual way, as on the plane, any Alpha graph contained in such an open set of the sphere. On an arbitrary surface, we will call an open set *disklike* if it is homeomorphic to an open disk of the plane. This is equivalent to being homeomorphic to the whole plane and, by a celebrated theorem by Riemann, to being homeomorphic to any non-empty simply connected open set of the plane.[70] There are enough disklike open sets on any surface.

Proposition 3.4. The topology of any surface has a base of disklike open sets.

Proof. Let O be an open set of any surface and let $p \in O$ be a point in it. Since some atlas covers the surface, let $\varphi : U \to V$ be a chart from an open set U of the surface onto an open subset V of the plane, such that $p \in U$. Then $O \cap U$ is an open set and its image $\varphi(O \cap U)$ is an open set of the plane, because φ is a homeomorphism. Since $p \in O \cap U$, we obtain $x = \varphi(p) \in \varphi(O \cap V)$ and there exists a radius $\epsilon > 0$ such that the open disk $D_\epsilon(x) = \{y \in \mathbb{R}^2 \mid |y - x| < \epsilon\}$ is contained in $\varphi(O \cap V)$.

---

[70] This was originally stated in Riemann 1851. For a current rendering, see Ahlfors 1979.

Now $\varphi^{-1}(D_\epsilon(x))$ is an open set of the surface that satisfies $p \in \varphi^{-1}(D_\epsilon(x)) \subseteq O$, and it is certainly disklike. □

Any Alpha graph contained in a disklike open set is mapped by a homeomorphism onto an Alpha graph on the plane. Therefore, we can read it with the usual interpretation of Alpha graphs on the plane, and we may apply the usual rules of transformation to it.

Theorem 3.9. Given a disklike open set on any surface, the system of all Alpha graphs contained in this open set, with the same interpretation and rules of transformation as on the plane, is a graphical version of classical propositional logic.

In this case we may also consider Beta graphs, modal Gamma graphs, and intuitionistic existential graphs, with the same constraint of being completely contained in the fixed disklike open set.

On the other hand, instead of limiting the surface that contains the Alpha graphs as we did on the sphere, we can also restrict the type of cuts as we did on the cylinder, the torus, and the Möbius strip. On an arbitrary surface, we call a Jordan curve *minimal* if its complement in the surface has *two* connected components, and exactly *one* of them is a disklike open set. On the plane, all Jordan curves are minimal, but on the sphere, none. On the cylinder, the torus, and the Möbius strip, the minimal Jordan curves are exactly the contractible ones. If we limit ourselves to Alpha graphs whose cuts are all minimal, then each cut has the disklike open set as its inside and the other component as its outside. For topological reasons, the exteriors of different cuts always have points in common, so there is only one outside for the whole graph, and the inside direction is consistent for different cuts. Therefore, we can read the graph with the same interpretation of the Alpha graphs on the plane and we can consistently define the parity of the areas. In order to apply all the usual rules of transformation, we have to make sure that every graph with only minimal cuts is contained in the disklike component of some minimal cut. In this case, we say that the surface *has enough minimal Jordan curves*.

Theorem 3.10. On a surface with enough minimal Jordan curves, the system of *minimal Alpha graphs*, composed of the Alpha graphs whose cuts are all minimal, with the same interpretation and rules of transformation as on the plane, is a graphical version of classical propositional logic.

With this constraint, we can also consider all other known systems of existential graphs on such a surface. This underlines once more the relevance of the inside direction in these graphs, because wherever that direction is consistently defined, we obtain essentially the same existential graphs as on the plane.

From a logical perspective, the real challenge remains to specify an interpretation and adequate rules of transformation for Alpha graphs with cuts for which it is not possible to consistently define an inside direction. An interesting particular case is given by those cuts which are naturally interpreted as a binary relation. From a geometric point of view, the question arises about the classification of possible Jordan curves on a given surface, which is very close to calculating its fundamental group. Furthermore, existential graphs pose the problem of classifying finite families of disjoint Jordan curves on surfaces. Finally, the problem remains about the further description of minimal Jordan curves and the surfaces with enough curves of this class. And so, existential graphs are expanding their abode far beyond their original natural environment.

Postscript

The abstract notion of a surface was clarified in the field of differential geometry.[71] The mathematical definition of Alpha graphs on surfaces is an immediate generalization of its formalization on the plane.[72] Existential graphs on nonplanar surfaces were introduced by Oostra[73] and are developed further in joint work with some of his students.[74] The theorem about Alpha graphs with contractible curves was first established on the torus.[75]

---

[71] See, for example, Malliavin 1972, do Carmo 1976, and Jost 2005.
[72] See the postscript in *Section 3.1*.
[73] Oostra 2019a.
[74] See Arana 2020, and Niño 2021.
[75] Arana 2020.

Fernando Zalamea
4 Around Arengas, Vargas, and Oostra Models for Peirce's Thought

Abstract: We highlight the achievements obtained in Chapters 1–3 of *Advances in Peircean Mathematics: The Colombian School*, and propose an integrated understanding of the results therein presented, thanks to the use of Riemann surfaces. A combination of semiotic, logic, category theory, complex variables, geometry, offers a simple model to encompass the diversity of Peirce's thought.

Keywords: Peirce; Riemann; mathematics; logic; category theory

Peirce's most appreciated work, either philosophical or semiotical, is deeply connected to his studies in *mathematics and logic*. The *continuity* of his system, interconnecting each of its parts, is an application of his more general *synechistic* standpoint: continuity, operative in Nature and Culture, turns out to be also operative along Peirce's own architectonics. A back-and-forth between global general (philosophical) concepts and local particular (mathematical) techniques is one of the main forces of Peirce's approach. In particular, Peirce was able to introduce very accurate ideas and techniques to support his main conjectures, around (1) the Pragmaticist Maxim, (2) the Continuum, and (3) the Existential Graphs. Precisely, *Chapters 1–3* in this compilation address these three main themes of Peirce's thought.

The architectonics of the system[1] interweaves some fundamental *local* and *global* techniques. If the central standpoint of Peirce's methodology is that we know by, and around, *signs*, yielding thus (1) the Pragmaticist Maxim, a natural perspective to undestand knowledge (and the world) becomes then a *continuity principle*, which allows signs to *contaminate* each other around borders, something captured along (2) Peirce's Continuum, and in turn reflected iconically in (3) the Existential Graphs (see *Figure 4.1*). There, the very layers of the architecture allow to go *back-and-forth* between the local (inscription in the graphs) and the global (continuum cultural irradiation).

---

[1] See *e.g.*, Murphey 1961, Parker 1998, or, Zalamea 2012a.

Fernando Zalamea, Fernando Zalamea (1959) is Professor at Universidad Nacional de Colombia.

https://doi.org/10.1515/9783110717631-004

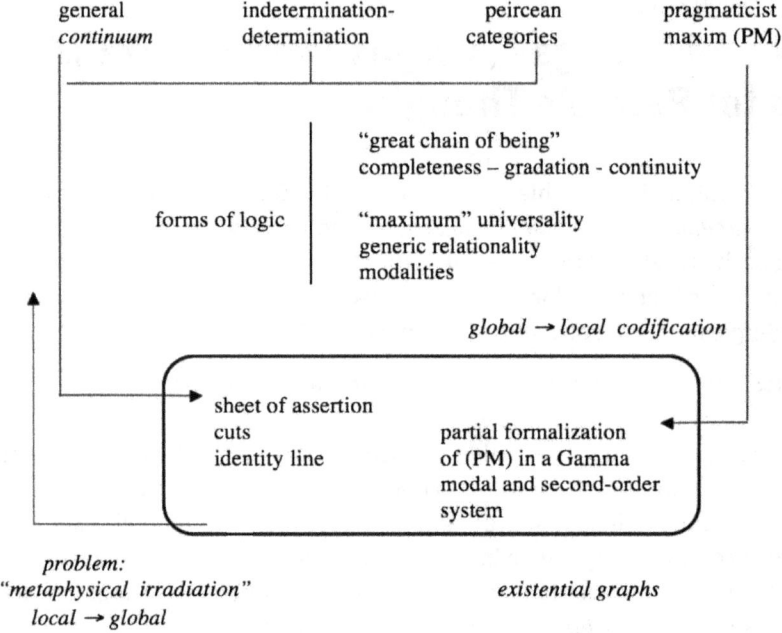

Figure 4.1: Global and Local Reflections in Peirce's Architectonics

A *back-and-forth* between generality and concreteness pervades Peirce's architectonics. Towards maximality, completeness, and universal continuity, some *global* forces fill the space of knowledge, in some *topological dense* sense. On another hand, an associated "density" of (global → local) codifications, and (local → global) irradiations, secures the adequate *complexity* of the many cognitive layers involved.

In what follows, we assess Arengas's *Chapter 1*, Vargas's *Chapter 2*, and Oostra's *Chapter 3*, and, building on them, we present their natural entanglements. Using Angie Hugueth's compact drawings for *Arengas's Gesture*, *Vargas's Gesture*, and *Oostra's Gesture*, we obtain a nice *gestural understanding* of the main bulk of their contributions. Finally, we speculate on a *higher-order topos approach* (Hugueth–Zalamea) to elucidate Peirce's architectonics, constructed on Arengas, Vargas, and Oostra mathematical modellings of (1) the Pragmaticist Maxim, (2) the Continuum, and (3) the Existential Graphs.

4.1 Arengas Models

The pragmatic maxim appears formulated several times throughout Peirce's intellectual development. The better known statement is from 1878, but more precise expressions appear (among others) in 1903 and 1905:

> Consider what effects which might conceivably have practical bearings we conceive the object of our conception to have. Then, our conception of these effects is the whole of our conception of the object. (**[1931–58]** CP 5.402; "How to Make Our Ideas Clear", 1878)
>
> Pragmatism is the principle that every theoretical judgement expressible in a sentence in the indicative mood is a confused form of thought whose only meaning, if it has any, lies in its tendency to enforce a corresponding practical maxim expressible as a conditional sentence having its apodosis in the imperative mood. (**[1931–58]** CP 5.18; "Harvard Lectures on Pragmatism", 1903)
>
> The entire intellectual purport of any symbol consists in the total of all general modes of rational conduct which, conditionally upon all the possible different circumstances, would ensue upon the acceptance of the symbol. (**[1931–58]** CP 5.438; "Issues of Pragmaticism", 1905)

The *Pragmaticist Maxim* (PM) (1903–1905, *possible* effects) is a modal extension of the pragmatic maxim (1878, *actual* effects), and signals that knowledge, seen as a semiotic-logical process, is pre-eminently contextual (versus absolute), relational (versus substantial), modal (versus determined), synthetic (versus analytic). According to Peirce's thought, we can only know through signs, and, according to the maxim, we can only know those signs through diverse correlations of its *conceivable* effects in interpretation contexts. The Pragmaticist Maxim "filters" the world by means of three complex webs which can "differentiate" the one into the many, and, conversely, can "integrate" the many into the one: a representational web, a relational web, a modal web. The pragmatic(ist) dimension emphasizes the correlation of all possible contexts: even if (PM) detects the fundamental importance of *local* interpretations, it also urges the reconstruction of *global* approaches, by means of appropriate relational and modal glueings of localities. A *diagrammatic scheme* of the Pragmaticist Maxim—which follows closely the 1903 and 1905 enunciations above stated—is the following[2] (see *Figure 4.2*):

---

[2] Zalamea 2012a, p. 55.

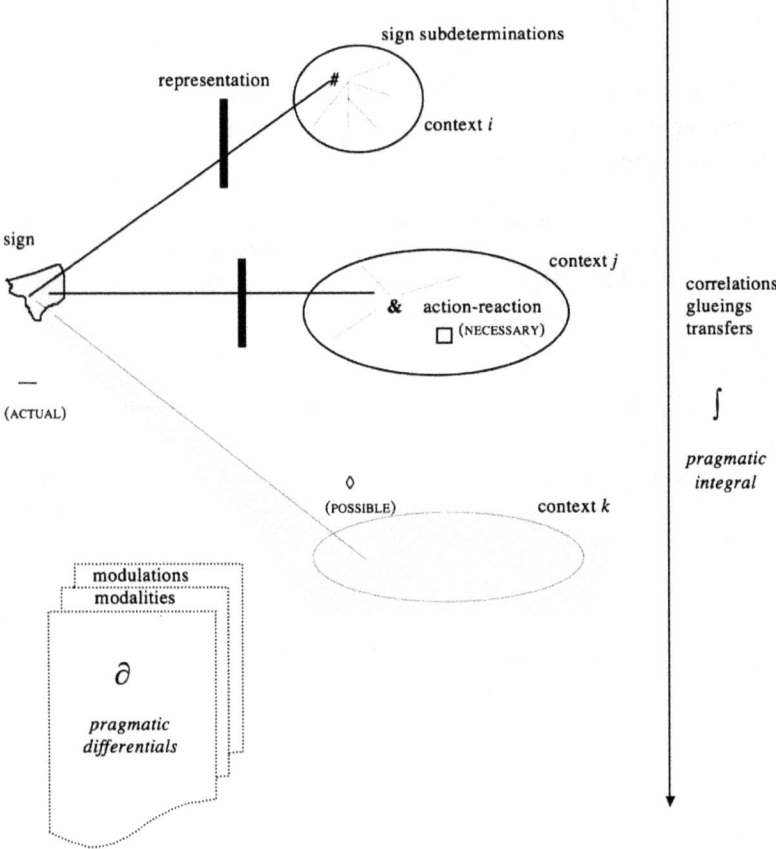

Figure 4.2: Peirce's Pragmaticist Maxim: "Differential and Integral Calculus"

In his PhD Thesis[3] (2016–2020), Gustavo Arengas has now *fully mathematized* (PM), as sketched in *Figure 4.2*, yielding a vast array of particular theorems in completely formalized contexts. Using the basic tools of Category Theory (synthetic morphism composition, instead of analytic elements decomposition), Arengas's dissertation studies the contextual, pragmatic understanding of a *given* sign (S) through its *visible* interpretation arrows (from the sign to its subdeterminations), hoping afterwards to capture back (S) thanks to the category-theoretic limit (L) of those arrows. This produces in a natural way an arrow $s : S \to L$, and the pragmaticist maxim can be encrypted by the simple property: (PM) *s is an isomor-*

---

[3] Arengas 2019.

phism.⁴ Thanks to the duality capabilities of Category Theory, the situation can be *inverted*, and a co-pragmaticist maxim (co-PM) expresses how a *hidden* sign (S) can emerge from its *residual* interpretations.⁵ The main bulk of Arengas's technical definitions, lemmas, and theorems, consists then in obtaining nice conditions in categories (generators, reflections, adjointness, separation, glueing, filtering, projectivity), in order to prove either (PM)⁶ or (co-PM).⁷ Applying those general theorems to the particular category of bi-Heyting Algebras, Arengas obtains (PM) representations in terms of orders and iterated modal operators.⁸ Further, the general (category-theoretic) and particular (order) results are studied through the lens of Voevodsky's Homotopy Type Theory,⁹ providing bounded, computable, proofs. In this way, Peirce's "proof of pragmaticism"¹⁰ turns out to be fully mathematized, and conclusively proved, at least in the very broad context of Category Theory.

On another hand, Arengas has now extended his Category Theory modellings of the Pragmaticist Maxim, lifting them to a *second-order* level of interpretations. Beyond a first-order look at objects and morphisms (isomorphisms and co-isomorphisms), Arengas has introduced *functors*, *adjoints*, and *monads* (see *Chapter 1, Section 4* above), to fully capture the deep dualities at play in the Pragmaticist Maxim. Arengas's basic idea is that the intrinsic *iteration* mechanism inscribed in a monad can model naturally the iterating *would-be* effects present in (PM) (and, dually, co-monad iterations can capture modal (co-PM) effects). From there, the general Eilenberg–Moore Category Theory calculus of monads provides a frame to use strong mathematical techniques (*e.g.*, Beck's Theorem, Sheaf Theory, Descent Theory, Beck–Chevalley Conditions), in order to model the various glueing, lifting, and projecting processes, hidden in (PM).

In this way, Arengas has opened up a vast new field of inquiry around the Pragmaticist Maxim. The many *layers* of his interpretations (first-order comparisons, and second-order stratifications) offer a fair, full, and faithful view of the complexities of knowledge. His mathematical finesse serves well Peirce's project to construct a "scientific metaphysics". Combining a *natural* Category Theory drive to understand the back-and-forth (*types and archetypes*) of mathematical thought, with Peirce's *natural* Synechistic Hypothesis to understand the world as a continu-

---

4 Arengas 2019, 13; see also *Chapter 1* above.
5 Arengas 2019, 14; see also *Chapter 1* above.
6 Proposition 1.2.17, Arengas 2019, p. 35; see also *Chapter 1* above.
7 Theorem 1.2.9, Arengas 2019, p. 29; see also *Chapter 1* above.
8 Theorem 2.2.4, Arengas 2019, p. 59.
9 Arengas 2019, pp. 97–147.
10 Robin 1997.

ous whole, Arengas's models offer many new insights to comprehend better ("scientific" approach) the overall architecture of Peirce's system ("metaphysical" bottoms). As we shall now see with Vargas's models for Peirce's Continuum, the possibility to envision completely the system (intertwining Pragmaticism and the Continuum) through modern and contemporary mathematical tools, enriches greatly the exactness and coherence of Peirce's views.

4.2 Vargas Models

Peirce's Continuum is one of his most original mathematical and logical contributions (the other one certainly being the Existential Graphs, see *Section 4.3* below). At the center of his philosophical, methodological, and semiotical edifice, the continuum pervades completely Peirce's architecture. Far from being "a castle in the air",[11] Peirce had very precise views on how to handle the continuum. Going well beyond Cantor's real numbers system, accepting infinitesimals, introducing strong reflexive and supermultitudinousness properties, delving into modalities, Peirce opened new paths to probe a *generic universal* continuum, of which Cantor's construction was to become just, in Peirce's words, "a first embryo of continuity".[12] The genesis of Peirce's Continuum is better understood when Peirce's ideas are contrasted with Cantor's, through an analytical/synthetical pendulum, which defines well their different approaches and helps to circumnavigate the *continua* involved. Comparing Cantor's definitional experiment, ordinals, and well-ordered stratifications, with Peirce's real abstraction, multitudes, and a welded continuum, Moore[13] shows how the two views can be seen as almost "orthogonal". On another hand, comparing Cantor's logico-analytical approach, logical surgery, and theological idealism, against Peirce's mathematico-synthetical perspectives, diagrammatic gestures, and metaphysical realism, Maddalena[14] emphasizes the power of "orthogonality" inscribed in the alternative continuum. Peirce's approach acquires a profound higher-dimensional meaning, which complements and completes our traditional understanding of continuity.

A century after Peirce, despite some attempts to capture partially some of the characteristics of Peirce's Continuum (Murphey, Herron, Havenel, Ehrlich, Zalamea), a *full global* model for it, which encompassed *all* of its generic/supermulti-

---

[11] Murphey 1961, p. 407.
[12] For a precise view of the situation, see Zalamea 2012a, pp. 3–10.
[13] Moore 2010, Moore 2015.
[14] Maddalena 2012.

tudinous, reflexive/inextensible, and modal/plastic characteristics[15] was still to be imagined, against all cutting prejudices of the "experts" involved (including this writer). As often happens in mathematics, the solution turned out to be as simple, as deep. Francisco Vargas[16] combined two powerful ideas to produce a straightforward ZFC model for Peirce's Continuum[17] something which seemed in principle very difficult, or almost impossible: (1) first, a *series* of copies of the real line is *iterated* along the class of *all* ordinals, (2) second, the *order* relation of set theoretic containment in the iterated model is *inverted* ("Order *E*"). The result produces an infinitely ordinal-iterated tree of real lines, with its branches looking *down* (via *E*). Through that extremely simple characteristic, any local, partial, cut in the tree ("Monad") turns out to be isomorphic to the global, whole, tree (six lines in the proof!) (for all this, see *Chapter 2, Section 2* above). From there, the main generic/supermultitudinous, reflexive/inextensible, properties of Peirce's Continuum are obtained at once, and with some more detail, modalities can also be defined and developed through ordinal levels and ramifications, tracing out the main properties of a modal Kripke model (see *Chapter 2, Section 3* above). There are no more points whatsoever, only extended parts, and a continuous weldedness governs the model, as predicted by Peirce.

In this volume, Vargas has developed further the nice *(i)* mereological *(Section 2.3)*, *(ii)* infinitesimal *(Sections 2.5 and 2.7)*, *(iii)* infinitarian *(Section 2.5)*, *(iv)* multidimensional *(Section 2.6)*, *(v)* algebraic *(Section 2.7)*, *(vi)* geometric *(Section 2.8)* and *(vii)* analytic *(Section 2.9)* properties of his model. It is extremely interesting that these multivalent approaches arise *naturally* from the very definition of the model: *(A)* extending the *usual* Kripke interpretation, sheaves emerge over a topological bottom on the Kripke construction (work still in progress by Vargas), *(B)* extending the *usual* decimal notation for Cantor's real numbers, transfinite ordinal representations for Peirce's Continuum numbers are obtained, *(C)* extending the *usual* Cantor correspondence between line and plane, a correspondence between "linear" Peirce's Continuum and its square shows again the difference between "cardinalities" and "dimensions", *(D)* extending the *usual* algebraic operations on $\mathbb{R}$, the algebraic operations on $\mathcal{C}_{Ord}$ become related to Cantor's Normal Form for ordinals, *(E)* extending the *usual* point-free properties of Leśniewski's mereological systems, both local ($\mathcal{C}_\alpha$) and global ($\mathcal{C}_{Ord}$) sections in Vargas's model verify easily the good properties of extended regions (Whitehead), beyond indivisible points (Cantor).

---

15 Zalamea 2012a, p. 53.
16 Vargas 2015.
17 For an English version of Vargas's 2012–2015 work, see Vargas and Moore 2021.

In all of these approaches, it is remarkable how a *natural fulfillment*, or inverse ordinal completeness, is achieved in Vargas's model, in a certain sense *maximizing the archetype of continuity*. From that maximal archetype, all other *continua types* are obtained through simple projections. This "archetype of continuity" will now be explored in the form of a further "archetype of logical thought", as Existential Graphs enter into the picture.

4.3 Oostra Models

We turn now to the *Existential Graphs* (EG), in Peirce's words, his "chef-d'oeuvre".[18] Peirce invented his Existential Graphs (circa 1896) to cover diagrammatically a wide diversity of logics, ranging from classical propositional calculus (Alpha Graphs), to classical first-order logic (Beta Graphs), to modal calculi (Gamma Graphs I), and second-order logic (Gamma Graphs II).[19] Over a blank sheet of assertion (representing Truth), cuts are made (representing Negation, via an Alpha language, and representing Possibilities, via a Gamma language), which divide the sheet in nested cut regions. Following *precise control rules*, one can introduce, eliminate, and transmit information around those nets (in particular, via a Beta language, one can extend, or restrict, a Line of Identity along the cuts, representing the Existential Quantifier: this is the reason for the generic name "Existential Graphs"). The *simultaneous* axiomatization of classical propositional calculus and purely relational first-order logic, with *five uniform generic rules* (double Alpha cuts, insertion, erasure, iteration and deiteration), renders explicit the technical *common roots* for both *calculi*, which have been entirely ignored in all other available presentations of classical logic. The *same rules* detect, in the context of Alpha language, the handling of classical negation and conjunction, and, in the context of Beta language, the handling of the existential quantifier: something just unimaginable for any logic student raised into Hilbert-type logic systems. (EG) show thus that there exists a kernel, a *"real general"* for classical thought, a kernel which, in some representational contexts, gives rise to the clas-

---

18 Letter to Jourdain, 1908, *cfr.* Roberts 1973, p. 110.
19 For the first exhaustive presentation of Peirce's writings on (EG), see [**2019–21**]. For the first complete and rigorous mathematical proofs around the Alpha, Beta, and Gamma formal systems, see Oostra 2018.

sical modes of connection, and which, in other contexts, gives rise to the classical modes of quantification.[20]

(EG) remained ignored for sixty years—in good part due to Quine's infamous review, where he qualified (EG) as some sort of "good entertainment"[21]—, until a remarkable situation emerged at the beginning of the 1960s, with Roberts's and Zeman's independent PhD dissertations on (EG).[22] Then, since the 1990s, (EG) came again into the picture, through the work of Burch,[23] Brady and Trimble,[24] Zalamea,[25] Pietarinen,[26] Sowa,[27] and other scholars, and many mathematical perspectives were offered to understand (EG): Combinatorial Topology, Topological Logic, Algebraic Logic, Nonstandard Logics, Category Theory, Graph Theory, Game Theory, Computer Science, Artificial Intelligence, Complex Variables, etc.

Nevertheless, special mention is due to what has certainly become the *center of the mathematical study* of (EG) in the world. Arnold Oostra's school, situated in Ibagué, Colombia, a far distant, abandoned academic place which reminds us of the isolation of Peirce's Arisbe, has produced in the last twenty years some twenty first-rate Undergraduate and Master dissertations on Peirce's logical work, from *modern* mathematical perspectives, with all the due rigour of a trained mathematician. The completeness proofs for Alpha, Beta, and Gamma systems (as well as an Alpha decision method[28]), have acquired a fully rigorous technical support through Oostra's school, something not entirely accomplished in Roberts's or Zeman's dissertations. On another hand, Oostra was the first to extend Peirce's graphs to a non-classical realm, constructing *new diagrams for Intuitionistic Logic* (IEG), adapting the inference rules, and proving completeness.[29] One of his main results shows that an intuitionistic (IEG) implication yields a classical (EG) implication,[30] but not conversely, revealing that the *true topological difference* between those logics consists, in diagrammatic terms, in allowing some sort of *separation*, but *not glueing* (see *Chapter 3, Section 2* above). The importance

---

20 For a general and vague discussion of the mathematical, methodological and philosophical aspects of (EG), *cfr.* Zalamea 2012a. For a particular and precise discussion, see Oostra's *Chapter 3* above.
21 Quine 1934, 553.
22 Roberts 1963, Zeman 1964.
23 Burch 1991.
24 Brady and Trimble 2000.
25 Zalamea 2012a (Spanish original edition, 2001).
26 Pietarinen 2006, Ma and Pietarinen 2018.
27 Sowa 2008, Sowa 2013.
28 Oostra 2016.
29 Oostra 2010, Oostra 2011.
30 Oostra 2010, pp. 49–50.

of (IEG) must be emphasized, since it provides the first deep *intrinsic natural* connection between Peirce's topo/logical ideas and Topology (the natural models for Intuitionism being topological spaces, according to Tarski).

Going beyond, Oostra has lately begun to explore *spatial extensions* of the (EG) (see *Chapter 3, Section 3* above), in non-planar elementary surfaces, as the cylinder, the sphere, or the torus,[31] showing a yet mysterious interdependence between the different *shapes* of the sheet of assertion and the different *logics* therein inscribed. The developments of (EG) in these two-dimensional Riemann surfaces, non-isomorphic to the plane (by their compactness or their genus > 1),[32] may be one of the most intriguing, and potentially fruitful developments of the *Logic of the Future*. Along these spatial developments, a profound blend of topological considerations, algebraic models, and logical rules, offers a fully Peircean methodology to study the *borders* of knowledge.

4.4 Hugueth Drawings

Angie Hugueth, a young artist and mathematician, has accomplished some telling drawings on Arengas's, Vargas's, and Oostra's works.[33] The first step has been to encrypt in a single, *simple gesture*, the main idea of each of the above contributions (*Chapters 1, 2, 3*, synthesized in turn in *Sections 4.1, 4.2, 4.3* above). Then, once the gestures have been proposed, the second step has been to draw compact diagrams on them.

---

31 Oostra 2019a, Oostra 2019b.
32 For a complex variable anticipation, see Zalamea 2010d.
33 Hugueth 2021.

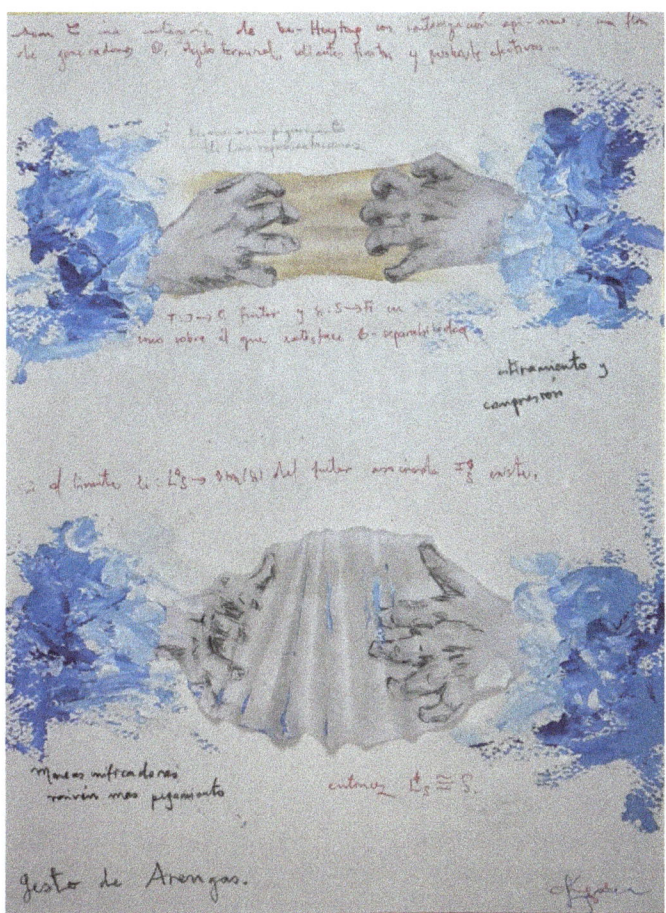

Figure 4.3: "Gesto de Arengas" (Hugueth 2021)

In this way, *Arengas's Gesture* (*Figure 4.3*) condenses the *back-and-forth stratifications* between a sign (S) and its pragmatic representation (L) (*Pragmaticist Maxim*, Chapter 1). Here, Hugueth uses two hands (representing S, L) which first try to reach each other (upper part of the diagram), and then become welded together through a sort of accordion ("estiramiento y compresión" / stretching and compression, yielding an isomorphism) which provides their natural intertwinement (lower part of the diagram). Hands emerge from deep waters (plasters of blue oil painting), trying to obtain some semiotic clearness, beyond the *chiaroscuro* understanding of the world.

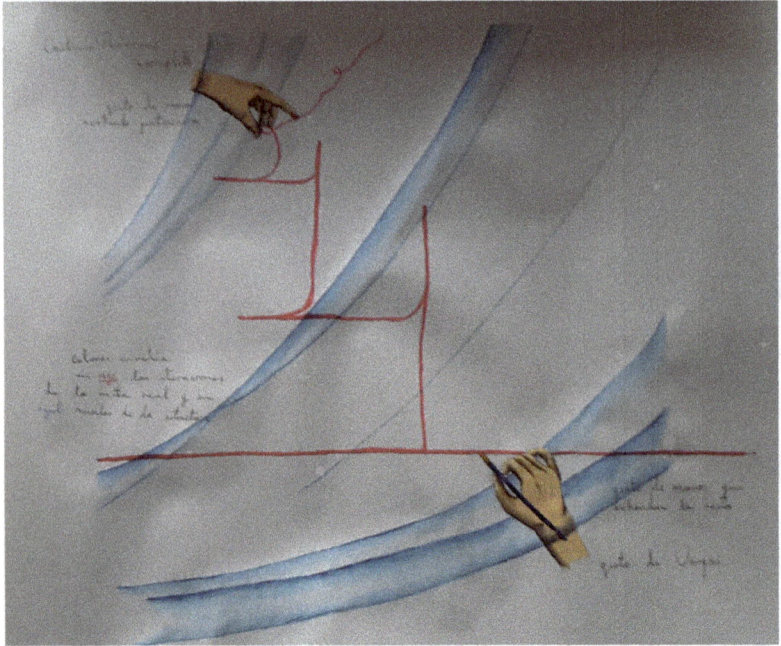

Figure 4.4: "Gesto de Vargas" (Hugueth 2017)

Vargas's Gesture (*Figure 4.4*) sums up the *infinite reflection* of fibers over an imaginary sheaf (*Peirce's Continuum, Chapter 2*). Again, moving and dynamic hands orient the viewer towards the complexity of knowledge, with all its different layers and mirrors. While red lines represent the fundamental doubling [fiber(fiber)] of the real line ("iteración" / iteration), developed towards infinity (*Ord*, whirling path, pointed by the upper indexing finger), some asymptotic circular curves (blue watercolors) open up the modal realm ("estructura" / structure), where all *possibilia* expand the spectrum of the continuum.

Oostra's Gesture (*Figure 4.5*) captures the essential separation and glueing processes associated to intuitionistic *logical iterations* (*Existential Graphs, Chapter 3*). Hugueth's hands express here the double movement (yellow arrows) of tearing apart and welding together Alpha cuts (red and blue pastel circles - "deformación, obstrucción" / deformation, obstruction), something related to the comparison (white border) between intuitionistic (IEG) implication and classical (EG) implication ("compactar, unificar" / to compactify and unify).

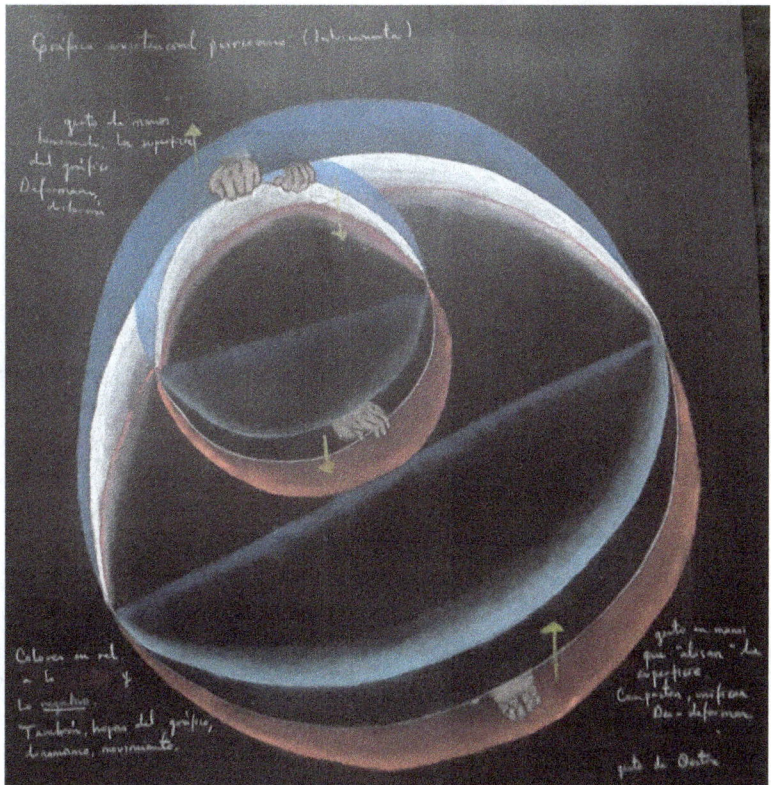

Figure 4.5: "Gesto de Oostra" (Hugueth 2017)

With these simple, but very imaginative drawings, Hugueth delves into the ingenuous, but ingenious, understanding that youth can provide, when looking to a situation from high perspectives devoided of restraining technicalities. In a profound sense, *beauty* leads the way, as it happens often with mathematics. Peirce's triadic system[34] is exalted: mathematics (1) becomes fully linked with logic (2.2.3) through esthetics (2.2.1).

---

[34] Kent 1987.

4.5 Towards a Mathematical Architectonics of Peirce's Architectonics

As we have seen in Arengas's *Chapter 1*, Vargas's *Chapter 2*, and Oostra's *Chapter 3*, condensed in *Sections 4.1, 4.2, 4.3* above, some main conceptual forces act as central pillars of Peirce's architectonics: (1) *stratification*, related to the Pragmaticist Maxim, (2) *reflection*, an essential ingredient of Peirce's Continuum, and (3) *iteration*, one of the main features of the Existential Graphs. In what follows, we concentrate on stratification, reflection, and iteration, and explore their many *mathematical intertwinements*, beyond some of their methodological and esthetical connections, as mentioned in *Section 4.4* thanks to Hugueth's drawings.

Category Theory provides an excellent environment to capture stratification (through category-theoretic layers, either internally, via free objects, or externally, via natural transformations), reflection (via adjoint functors), and iteration (via monads).[35] Another nice *milieu* is the theory of *Functions of a Complex Variable*, thanks to the stratified hierarchy of holomorphic and meromorphic functions, their reflections around the Riemann–Roch theorem, and iterations related to analytic continuation. *Topology* is a third natural environment, where stratification (derived sets), reflection (fixed point theorems), and iteration (sequences or filters), are common techniques used to understand limiting processes. Finally, *Abstract Algebra*, with Heyting (co-)algebras or Kripke models for non-classical logics, offers a context where stratification (intermediate connectives), reflection (intermediate axioms), and iteration (accessibility relations) play some major roles.

A fundamental *open problem* to understand Peirce's architectonics from a mathematical point of view, consists in *blending naturally* together the four perspectives just mentioned: (A) Category Theory, (B) Complex Variables, (C) Topology, (D) Algebra. A good *laboratory* for this inquiry lies in the *spatial extensions of Existential Graphs* (see Oostra's *Chapter 3, Section 3* above, or the end of our *Section 4.3* above). There, beyond Oostra's focused studies on the sphere and the torus, a hopeful path consists in Hugueth's Undergraduate Thesis, *Topos of Existential Graphs over Riemann Surfaces*.[36] On one hand, themes (B) and (C) are condensed in Riemann Surfaces, leading to some *Main Conjectures*: *(i)* all (EG) *local* logics over a Riemann Surface are super-intuitionistic (Oostra), and most of them classical (Hugueth), *(ii)* the *global* logic over a sphere is non-classical (Oostra),

---

35 For a detailed account, see Zalamea 2010a.
36 Hugueth 2022.

(iii) the *global* logics over Riemann Surfaces of genus $n \geq 1$ are non-classical, with at least $3n$ notions of negation (some of them paraconsistent) (Zalamea–Hugueth). On the other hand, themes (A) and (D) are to be developed, understanding, first, *a given* (EG) as a presheaf,[37] viewing, second, *all* (EG)s as a Grothendieck topos, and, delving, third, into the algebraic bottom of (EG)s via properties of the *classifier object* in the sheaf topos. Through these new perspectives, stratification, reflection, and iteration, become naturally intertwined in the many mirroring adjunctive layers of the topos.

Guerino Mazzola has formalized a sophisticated *Gesture Theory*,[38] coming from questions related to the crucial role of *improvisation* in music. Therein, *gestures* are understood as embodiments of graph skeletons in suitable topological spaces, while *hypergestures* are defined to capture the possibility of the fundamental *iteration* "gestures on gestures". Hypergestures offer thus a deep geometric understanding of *spaces of gestures*, through function spaces, or exponentials, in a topos. Going further, Juan Sebastián Arias has introduced gestures on *locales* (axiomatic presentations of Heyting complete algebras) and developed geometric realization approaches to abstract gestures.[39] Using Mazzola and Arias's methodologies we can delve a little further on *Hugueth's gestures*. In fact, if we superpose (in a sort of four-folded Riemann Surface, akin to $\sqrt[4]{z}$) *Arengas's gesture* (Figure 4.3 above), *Vargas's gesture* (Figure 4.4), and *Oostra's gesture* (Figure 4.5), over Zalamea's work on *Peirce's logic of continuity*, we obtain a *hypergesture* (gesture on gestures) which condenses the *borders, limits, and ends*[40] of Peirce's architectonics (see *Figure 4.6*).

---

37 Gangle, Caterina, and Tohme 2020.
38 Mazzola 2018.
39 Arias 2017, *Chapters 2, 3*. After studying the mathematical behavior of gestures in topoi, 2-categories, and internal categories (*Chapters 4, 5*), Arias presents at the end of his dissertation a "Philosophical Framework for Gesture Theory" (*Chapter 6*), where he builds on some central semiotical ideas of Peirce. In this way, Arias dissertation may be understood as a tangential part of the Colombian School around *Advances in Peircean Mathematics*.
40 *Cfr.* Arias 2017, *Chapter 3*.

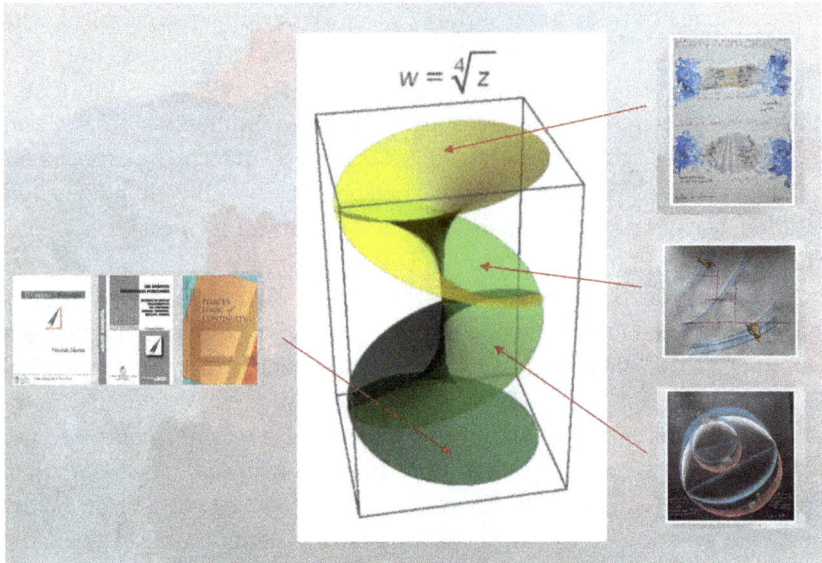

Figure 4.6: A "Riemann Surface" View of Peirce's Architectonics

In this "compact" Hugueth view (which reflects the compactness of the Riemann Surface of $\sqrt[4]{z}$), Arengas's "accordion" hands, Vargas's "autoreflexive" hands, and Oostra's "separating/glueing" hands (see *Section 4.4* above), provide a *natural gestural* dialogue between stratification, reflection, and iteration. The accordion synthesizes the never ending back-and-forth stratification between borders of knowledge, the autoreflexive monad captures the reflection properties between part and whole, the pulling topo/logical hands encrypt the non-classical richness of discrete/continuous iterations.

Finally, we come back to the main global and local reflections in Peirce's architectonics (see *Figure 4.1* above). If we apply Peirce's *Pragmaticist Maxim* (*Figure 4.2*) to the very sign/concept *Continuity*,[41] Vargas's model comes to be *ramified* at all levels of the architecture. For example, at level 2.1 (Phenomenology) of Peirce's Classification of Sciences,[42] the good reflexive and modal properties of the model help to explain better the sometimes obscure *prescisive layered* characteristics of Peirce's three cenopythagorean categories. On the other hand, if we apply further the Maxim to the sign/concept *Existential Graphs*,[43] Oostra's models

---

41 For details, see Zalamea 2012a, *Chapters 3, 4*.
42 *Cfr.* Kent 1987.
43 For details, see Zalamea and Nubiola 2011.

help us to navigate the rich spectrum of *contextual* logic, algebraic, and geometric pragmaticist interpretations, condensed in the *unique archetypical* axiomatics of (EG) iteration and its *multivalent type* embodiments.

Weaving together the many insights of Arengas, Vargas, Oostra, and Hugueth, we obtain a solid mathematical, methodological, logical, topological, esthetical, semiotical, and philosophical *elastic structure*. This shows how Peirce's "castle in the air" (Murphey), far from being just a kind of "good entertainment" (Quine), has now acquired the needed technical supports to be able to fly plastically in the skies.

Peirce Bibliography

Peirce, Charles Sanders (**2010**). *Philosophy of Mathematics: Selected Writings*. Ed. by Matthew Moore. Indiana University Press.
Peirce, Charles Sanders (**1906**). "Prolegomena to an Apology for Pragmaticism". In: *The Monist* 16.4, pp. 492–546.
Peirce, Charles Sanders (**1892**). "The Law of Mind". In: *The Monist* 2.4, pp. 533–559.
Peirce, Charles Sanders (**1993**). *Reasoning and the Logic of Things (Cambridge Lectures 1898)*. Ed. by K. L. Ketner. Cambridge: Harvard University Press.
Peirce, Charles Sanders (**1976**). *The New Elements of Mathematics*. Ed. by Carolyn Eisele. Vol. 1–4. The Hague: Mouton.
Peirce, Charles Sanders (**2019–21**). *Logic of the Future*. Ed. by Ahti-Veikko Pietarinen. Vol. 1–3. Peirceana. Berlin: De Gruyter.
Peirce, Charles Sanders (**1931–58**). *Collected Papers of Charles Sanders Peirce*. Ed. by C. Hartshorne, P. Weiss, and A. W. Burks. Vol. 1–8. Cambridge: Harvard University Press.
Peirce, Charles Sanders (**1999–98**). *The Essential Peirce*. Ed. by N. Houser and et.al. Bloomington: Indiana University Press University Press.
Peirce, Charles Sanders (**1982–2009**). *Writings of Charles S. Peirce. A Chronological Edition*. Ed. by E. C. Moore et al. Vol. 1–6, 8. Bloomington: Indiana University Press.

Secondary Bibliography

Acosta, Jennifer A. and Margarita Jiménez (2010). *El método de decisión Alfa*. Undergraduate thesis. Ibagué (Colombia): Universidad del Tolima.
Acosta, Lorenzo (2015). *Temas de teoría de retículos*. Bogotá: Universidad Nacional de Colombia.
Ahlfors, Lars V. (1979). *Complex Analysis*. Third edition. New York: McGraw-Hill.
Altenkirch, Thorsten and Alexander Green (2010). "The Quantum IO Monad". In: *Semantic Techniques in Quantum Computation*. Ed. by Simon Gay and Ian Mackie. Cambridge: Cambridge University Press, pp. 173–205.
Arana, Camilo (2020). *Gráficos existenciales Alfa de Peirce sobre el toro*. Undergraduate thesis. Bogotá: Universidad Nacional de Colombia.
Arengas, Gustavo (2019). "La máxima pragmática peirceana: modelos categóricos, dualización, aproximaciones algebraicas y modalizaciones lógicas". PhD thesis. Bogotá: Universidad Nacional de Colombia.
Arias, Juan Sebastián (2017). "Gesture Theory: Topos-Theoretic Perspectives and Philosophical Framework". PhD thesis. Bogotá: Universidad Nacional de Colombia.
Baez, John and Mike Stay (2011). "Physics, Topology, Logic and Computation: A Rosetta Stone". In: *New Structures for Physics*. Ed. by Bob Coecke. Berlin: Springer-Verlag, pp. 95–172.
Barwise, Jon and Solomon Feferman (1985). *Model-Theoretic Logics*. Vol. 8. Perspectives in Logic. New York: Springer-Verlag.
Beck, Jonathan (1967). "Triples, algebras and cohomology". PhD thesis. New York: Columbia University.
Bell, John L. (1998). *A Primer of Infinitesimal Analysis*. Cambridge: Cambridge University Press.
Bénabou, Jean and Jacques Roubaud (1970). "Monades et descente". In: *C. R. Acad. Sc. Paris* t.270, pp. 96–98.
Blyth, Thomas S. (2005). *Lattices and Ordered Algebraic Structures*. London: Springer.
Brady, Geraldine and Todd H. Trimble (2000). "A Categorical Interpretation of C. S. Peirce's Propositional Logic Alpha". In: *Journal of Pure and Applied Algebra* 149, pp. 213–239.
Brady, Geraldine and Todd H. Trimble (n.d.). "A String Diagram Calculus for Predicate Logic and C. S. Peirce's System Beta (2000)".
Brioschi, Maria Regina (2020). *Creativity Between Experience and Cosmos. C.S. Peirce and A.N. Whitehead on Novelty*. Vol. 6. Whitehead Studien. Verlag Karl Alber.
Bunt, Harry and William Black (2000). "The ABC of Computational Pragmatics". In: *Abduction, Belief and Context in Dialogue. Studies in Computational Pragmatics*. Ed. by Harry Bunt and William Black. Amsterdam: Benjamins, pp. 1–46.
Burch, Robert (1991). *A Peircean Reduction Thesis. The Foundations of Topological Logic*. Lubbock: Texas Tech University Press.
Caicedo, Xavier (1990). *Elementos de lógica y calculabilidad*. Bogotá: Una empresa docente.
Caicedo, Xavier (1995). "Lógica de los haces de estructuras". In: *Rev. Acad. Colomb. Cienc.* 19.74, pp. 569–586.
Calderón, Fauner and Mabel Calderón (2021). *Gráficos Alfa para la lógica implicativa*. Undergraduate thesis. Ibagué (Colombia): Universidad del Tolima.
Cantor, Georg (1915). "Beiträge zur Begründung der transfiniten Mengenlehre (1895)". In: *Contributions to the Founding of the Theory of Transfinite Numbers*. Translation. New York: Dover.

https://doi.org/10.1515/9783110717631-006

Castillo, Mauricio (2009). *Lógicas implicativas y sus álgebras*. Undergraduate thesis. Ibagué (Colombia): Universidad del Tolima.
Castillo, Mauricio and Arnold Oostra (2010). "Álgebras para la lógica implicativa con conjunción". In: *Matemáticas: Enseñanza Universitaria* 18.2, pp. 31–50.
Castro, Geraldine (2021). *Geometría plana con gráficos existenciales*. Undergraduate thesis. Ibagué (Colombia): Universidad del Tolima.
Chagrov, Alexander and Michael Zakharyaschev (1997). *Modal Logic*. Oxford: Clarendon Press.
Chellas, Brian F. (1980). *Modal Logic. An Introduction*. Cambridge: Cambridge University Press.
Corfield, David (2020). *Modal Homotopy Type Theory. The Prospect of a New Logic for Philosophy*. Oxford: Oxford University Press.
Dauben, Joseph W. (1982). "Peirce's place in mathematics". In: *Historia Mathematica* 9.3, pp. 311–325.
de Souza, Clarisse Sieckenius (2005). *The Semiotic Engineering of Human-Computer Interaction*. Cambridge: MIT Press.
De Tienne, André (2015). "The Flow of Time and the Flow of Signs: A Basis for Peirce's Cosmosemiotics". In: *The American Journal of Semiotics* 31, pp. 29–53.
Díaz, Daniela (2016). *Álgebras booleanas libres y gráficos Alfa*. Undergraduate thesis. Ibagué (Colombia): Universidad del Tolima.
do Carmo, Manfredo P. (1976). *Differential Geometry of Curves and Surfaces*. Englewood Cliffs (NJ): Prentice-Hall.
Ehrlich, Philip (2010). "The Absolute Arithmetic Continuum and its Peircean Counterpart". In: *New Essays on Peirce's Mathematical Philosophy*. Ed. by Matthew E. Moore. Chicago: Open Court, pp. 235–281.
Ehrlich, Philip (2021). "Contemporary Infinitesimalist Theories of Continua and Their Late Nineteenth- and Early Twentieth-Century Forerunners". In: *The History of Continua. Philosophical and Mathematical Perspectives*. Ed. by Stewart Shapiro and Geoffrey Hellman. Oxford University Press, pp. 328–346.
Fakir, Sabah (1970). "Monade idempotente associée à une monade". In: *C. R. Acad. Sci. Paris* t.270, pp. 99–101.
Florensky, Pavel (1997). *The Pillar and Ground of the Truth*. Princeton: Princeton University Press.
Fuentes, Camilo (2014). *Cálculo de secuentes y gráficos existenciales Alfa: Dos estructuras equivalentes para la lógica proposicional*. Undergraduate thesis. Ibagué (Colombia): Universidad del Tolima.
Gabbay, Dov M. and Rudolf Kruse, eds. (2000). *Abductive Reasoning and Learning*. Dordrecht: Springer-Science+Business Media.
Gangle, R., G. Caterina, and F. Tohme (2020). "A Generic Figures Reconstruction of Peirce's Existential Graphs (Alpha)". In: *Erkenntnis* https://doi.org/10.1007/s10670-019-00211-5.
Gerla, Giangiacomo (2021). "Point-Free Continuum". In: *The History of Continua. Philosophical and Mathematical Perspectives*. Ed. by Stewart Shapiro and Geoffrey Hellman. Oxford University Press, pp. 427–475.
Gödel, Kurt (1932). "Zum intuitionistischen Aussagenkalkül". In: *Anz. Akad. Wiss. Wien* 69, pp. 65–66.
Goldblatt, Robert (1979). *Topoi. The Categorial Analysis of Logic*. Amsterdam: North-Holland.
Goldblatt, Robert (1998). *Lectures on the Hyperreals. An Introduction to Nonstandard Analysis*. Vol. 188. Graduate Texts in Mathematics. Springer-Verlag.

Gómez, Andrea Y. (2013). *Gráficos Alfa para la lógica implicativa con conjunción*. Undergraduate thesis. Ibagué (Colombia): Universidad del Tolima.

Grothendieck, Alexander (1959). "Technique de descente et théorèmes d'existence en géométrie algébrique. I. Généralités. Descente par morphismes fidèlement plats". In: *Séminaire Bourbaki* 5. Exp. 190, pp. 299–327.

Grothendieck, Alexander and Michèle Raynaud (1971). *Revêtements étales et groupe fondamental (SGA 1)*. Paris: Société Mathématique de France.

Guerra, Ana María (2021). *Reglas nuevas para los gráficos Gama*. Undergraduate thesis. Ibagué (Colombia): Universidad del Tolima.

Hahn, Hans (1907). "Über die nichtarchimedischen Grössensysteme". In: *Sitzungsberichte der Kaiserlichen Akademie der Wissenschaften, Mathematisch-Naturwissenschaftliche Klasse (Wien)* 116, pp. 601–655.

Havenel, Jérôme (2008). "Peirce's Clarifications of Continuity". In: *Transactions of the Charles S. Peirce Society* 44.1, pp. 86–133.

Heyting, Arend (1930). "Die formalen Regeln der intuitionistischen Logik". In: *Sitzungsber. preuss. Akad. Wiss. Berlin*, pp. 42–71, 158–169.

Heyting, Arend (1971). *Intuitionism. An Introduction*. Amsterdam: North-Holland.

Hugueth, Angie (2021). *Three Simple Gestures which Capture a Mathematical Understanding of C. S. Peirce's System*. Tech. rep. Bogotá: Universidad Nacional de Colombia.

Hugueth, Angie (2022). *Topos of Existential Graphs over Riemann Surfaces*. Undergraduate thesis. Bogotá: Universidad Nacional de Colombia.

Janelidze, George and Walter Tholen (1994). "Facets of Descent I". In: *Applied Categorical Structures* 2, pp. 245–281.

Jech, Thomas (1997). *Set Theory*. Second edition. Berlin: Springer-Verlag.

Jost, Jürgen (2005). *Riemannian Geometry and Geometric Analysis*. Fourth edition. Berlin: Springer.

Joyal, André (2002). "Quasi-categories and Kan complexes". In: *J. Pure Appl. Algebra* 175, pp. 207–222.

Kelly, Max and William Lawvere (1988). "On the Complete Lattice of Essential Localizations". In: *Bull. Société Mathematique de Belgique* XLI, pp. 289–319.

Kent, Beverley (1987). *Charles S. Peirce. Logic and the Classification of Sciences*. Montréal: McGill - Queen's University Press.

Ketner, Kenneth Laine and Hilary Putnam (1992). "Introduction: The Consequences of Mathematics". In: *Reasoning and the Logic of Things*. Ed. by Kenneth Laine Ketner. Cambridge: Harvard University Press, pp. 1–54.

Kunen, Kenneth (1980). *Set Theory. An Introduction to Independence Proofs*. Amsterdam: North-Holland.

Largeault, Jean (1993). *Intuition et Intuitionisme*. Paris: Vrin.

Lautman, Albert (2011). *Ensayos sobre la dialéctica, estructura y unidad de las matemáticas modernas*. Bogotá: Universidad Nacional de Colombia.

Lawvere, William (1962). *The category of probabilistic mappings*. Manuscript. URL: https://ncatlab.org/nlab/files/lawvereprobability1962.pdf.

Lawvere, William (1989). "Display of graphics and their applications, as exemplified by 2-categories and the Hegelian 'taco'". In: *Proceedings of the First International Conference on Algebraic Methodology and Software Technology*. Iowa City: University of Iowa, pp. 51–74.

Lawvere, William (1994). "Cohesive Toposes and Cantor's *lauter Einsen*". In: *Philosophia Mathematica* 2.3, pp. 5–15.
Lawvere, William (1996). "Unity and Identity of Opposites in Calculus and Physics". In: *App. Cat. Struc.* 4, pp. 167–174.
Lawvere, William and Matías Menni (2015). "Internal choice holds in the discrete part of any cohesive topos satisfying stable connected codiscreteness". In: *TAC* 30.26, pp. 909–932.
Lesniewski, Stanisław (1916). *Foundations of the General Theory of Sets, I*. Moscow.
Levi-Civita, Tullio (1893). *Sugli infiniti ed infinitesimi attuali, quali elementi analitici*. Venezia: Ferrari.
López, Katherine (2013). *Modelos de Kripke para lógicas modales*. Undergraduate thesis. Ibagué (Colombia): Universidad del Tolima.
Lurie, Jacob (2009). *Higher Topos Theory*. Princeton: Princeton University Press.
Lurie, Jacob (2017). *Higher Algebra*. Preprint. URL: http://people.math.harvard.edu/~lurie/papers/HA.pdf.
Lützen, Jesper (2021). "Hjelmslev's geometry of reality". In: *Historia Mathematica* 54, pp. 95–116.
Ma, Minghui and Ahti-Veikko Pietarinen (2018). "Gamma Graph Calculi for Modal Logics". In: *Synthese* 195, pp. 3621–3650.
Mac Lane, Saunders and Ieke Moerdijk (1992). *Sheaves in Geometry and Logic. A First Introduction to Topos Theory*. New York: Springer-Verlag.
Maddalena, Giovanni (2012). "Peirce's Incomplete Synthetic Turn". In: *The Review of Metaphysics* 65.3, pp. 613–640.
Maddalena, Giovanni (2015). *The Philosophy of Gesture. Completing Pragmatists' Incomplete Revolution*. Montreal: McGill-Queen's University Press.
Malliavin, Paul (1972). *Géométrie différentielle intrinsèque*. Paris: Hermann.
Martín, Alejandro (2000). "Peirce y los modelos matemáticos del continuo". In: *Actas del III Congreso de la Sociedad de Lógica, Metodología y Filosofía de la Ciencia en España*. Departamento de Filosofía, pp. 51–60.
Martínez, Yorladys (2014). *Un modelo real para los gráficos Alfa*. Undergraduate thesis. Ibagué (Colombia): Universidad del Tolima.
Mazzola, Guerino (2018). *The Topos of Music*. Second edition. Vol. 1–4. Vol. 3 (*Gesture Theory*). New York: Springer.
Moggi, Eugenio (1989). "Computational lambda-calculus and monads". In: *Proceedings of the Fourth Annual Symposium on Logic in Computer Science*, pp. 14–23.
Molina, Fabián A. (2001). *Correspondencia entre algunos sistemas de lógica modal y los gráficos existenciales Gama de Peirce*. Undergraduate thesis. Ibagué (Colombia): Universidad del Tolima.
Montealegre, Luis E. (2020). *Los silogismos aristotélicos y los gráficos existenciales*. Undergraduate thesis. Ibagué (Colombia): Universidad del Tolima.
Moore, Matthew E. (2010). "Peirce's Cantor". In: *New Essays on Peirce's Mathematical Philosophy*. Ed. by Matthew E. Moore. Chicago: Open Court, pp. 323–362.
Moore, Matthew E. (2015). "Peirce's Prepunctual Continuum". In: *Cuadernos de Sistemática Peirceana* 7, pp. 127–138.
Moreno, John F. (2014). *Lógicas intermedias*. Undergraduate thesis. Ibagué (Colombia): Universidad del Tolima.
Murphey, Murray (1961). *The Development of Peirce's Philosophy*. Cambridge: Harvard University Press.

Myrvold, Wayne C. (1995). "Peirce on Cantor's Paradox and the Continuum". In: *Transactions of the Charles S. Peirce Society* 31.3, pp. 508–541.
Niño, Luisa F. (2021). "Gráficos existenciales Alfa sobre la esfera". Master's thesis. Ibagué (Colombia): Universidad del Tolima.
nLab authors (Aug. 2021a). *abductive reasoning*. URL: https://ncatlab.org/nlab/show/abductive+reasoning.
nLab authors (Aug. 2021b). *Science of Logic*. URL: https://ncatlab.org/nlab/show/Science+of+Logic.
Nubiola, Jaime (2008). "Charles S. Peirce (1839–1914)". In: *Handbook of Whiteheadian Process Thought*. Ed. by Michel Weber and Will Desmond. Vol. 2. De Gruyter, pp. 481–487.
Obradovic, Jovana (2016). *The Bénabou-Roubaud monadic descent result via string diagrams*. Preprint.
Oostra, Arnold (1997). "Álgebras de Heyting". In: Memorias del XIV Coloquio Distrital de Matemáticas y Estadística.
Oostra, Arnold (2010). "Los gráficos Alfa de Peirce aplicados a la lógica intuicionista". In: *Cuadernos de Sistemática Peirceana* 2, pp. 25–60.
Oostra, Arnold (2011). "Gráficos existenciales Beta intuicionistas". In: *Cuadernos de Sistemática Peirceana* 3, pp. 53–78.
Oostra, Arnold (2012). "Los gráficos existenciales Gama aplicados a algunas lógicas modales intuicionistas". In: *Cuadernos de Sistemática Peirceana* 4, pp. 27–50.
Oostra, Arnold (2016). "Peirce's Decision Method for Alpha Graphs Revisited". In: *Cuadernos de Sistemática Peirceana* 8, pp. 119–135.
Oostra, Arnold (2018). *Notas de lógica matemática*. Preprint. Ibagué: Universidad del Tolima.
Oostra, Arnold (2019a). "Existential Graphs on Nonplanar Surfaces". In: *Revista Colombiana de Matemáticas* 53.2, pp. 205–219.
Oostra, Arnold (2019b). "Representación compleja de los gráficos Alfa para la lógica implicativa con conjunción". In: *Boletín de Matemáticas* 26.1, pp. 31–50.
Oostra, Arnold (2021). "Equivalence Proof for Intuitionistic Existential Alpha Graphs". In: *Diagrammatic Representation and Inference*. Ed. by Amrita Basu *et. al*. Vol. 12909. Lecture Notes in Artificial Intelligence. Cham: Springer, pp. 188–195.
Oostra, Arnold and Daniela Díaz (2016). "Álgebras booleanas libres en álgebra, topología y lógica". In: *Boletín de Matemáticas* 23.2, pp. 143–163.
Orchard, Dominic (2014). "Programming contextual computations". PhD thesis. Cambridge: University of Cambridge.
Ortiz, Jorge and Juan Segura (2018). *Gráficos Alfa intuicionistas*. Undergraduate thesis. Ibagué (Colombia): Universidad del Tolima.
Parker, Kelly (1998). *The Continuity of Peirce's Thought*. Nashville: Vanderbilt University Press.
Peirce, Benjamin (1837). *An Elementary Treatise on Plane and Solid Geometry*. Boston: J. Munroe and Company.
Peirce, Charles Sanders (1992). *Reasoning and the Logic of Things*. Edited by Kenneth Laine Ketner, with introduction by Kenneth Laine Ketner and Hilary Putnam. Cambridge: Harvard University Press.
Petricek, Tomas (2017). "Context-aware programming languages". PhD thesis. Cambridge: University of Cambridge.
Pietarinen, Ahti-Veikko (2006). *Signs of Logic. Peircean Themes on the Philosophy of Language, Games, and Communication*. Dordrecht: Springer.

Pietarinen, Ahti-Veikko (2016). "On the Supreme Beauty of Logical Graphs". In: *Cuadernos de Sistemática Peirceana* 8, pp. 5–40.
Poveda, Yuri Alexander (2000). "Los gráficos existenciales de Peirce en los sistemas Alfa^0 y Alfa^{00}". In: *Boletín de Matemáticas* 7.1, pp. 5–17.
Prada, Ricardo (2012). *Gráficos existenciales Alfa y teoría de categorías*. Undergraduate thesis. Ibagué (Colombia): Universidad del Tolima.
Prada, Ricardo (2018). "Gráficos existenciales Gama, modelos de Kripke y haces". Master's thesis. Ibagué (Colombia): Universidad del Tolima.
Priest, Graham (2008). *An Introduction to Non-Classical Logic. From If to Is*. Cambridge: Cambridge University Press.
Quine, W.V.O. (1934). "Review - *Collected Papers* of Charles Sanders Peirce, Volume IV". In: *Isis* XXII, pp. 551–553.
Reyes, Gonzalo and Marek Zawadowski (1993). "Formal Systems for Modal Operators on Locales". In: *Studia Logica* 52.4, pp. 595–613.
Riemann, Bernhard (1851). "Grundlagen für eine allgemeine Theorie der Functionen einer veränderlichen complexen Grösse". Inauguraldissertation. Göttingen: Georg-August-Universität.
Roberts, Don D. (1963). "The Existential Graphs of Charles S. Peirce". PhD thesis. Urbana-Champaign: University of Illinois.
Roberts, Don D. (1973). *The Existential Graphs of Charles S. Peirce*. The Hague: Mouton.
Roberts, Don D. (1997). "A Decision Method for Existential Graphs". In: *Studies in the Logic of Charles Sanders Peirce*. Ed. by Nathan Houser, Don D. Roberts, and James Van Evra. Bloomington and Indianapolis: Indiana University Press, pp. 387–401.
Robin, Richard (1997). "Classical Pragmatism and Pragmatism's Proof". In: *The Rule of Reason. The Philosophy of Charles Sanders Peirce*. Ed. by Jacqueline Brunning and Paul Forster. Toronto: University of Toronto Press, pp. 145–146.
Robinson, Abraham (1974). *Non-standard Analysis*. Amsterdam: North-Holland.
Rueda, Ricardo (2011). *Matemáticas básicas con gráficos existenciales Beta*. Undergraduate thesis. Ibagué (Colombia): Universidad del Tolima.
Schreiber, Urs (2016). *Modern Physics formalized in Modal Homotopy Type Theory*. Preprint. URL: https://ncatlab.org/schreiber/files/MPfiMHTT200618.pdf.
Sierpiński, Wacław (1958). *Cardinal and Ordinal Numbers*. Warszawa: PWN - Polish Scientific Publishers.
Sowa, John F. (2008). "Conceptual Graphs". In: *Handbook of Knowledge Representation*. Ed. by F. van Harmelen. Amsterdam: Elsevier, pp. 213–237.
Sowa, John F. (2013). "From Existential Graphs to Conceptual Graphs". In: *International Journal of Conceptual Structures* 1, pp. 39–72.
Sowa, John F. (2020). *Peirce, Polya, and Euclid. Integrating Logic, Heuristics, and Geometry*. http://www.jfsowa.com/talks/ppe.pdf. Slides.
Taboada, Jorge and Danilo Rodríguez (2010). *Una demostración de la equivalencia entre los gráficos Alfa y la lógica proposicional*. Undergraduate thesis. Ibagué (Colombia): Universidad del Tolima.
Tall, David (1979). "The Calculus of Leibniz – An Alternative Modern Approach". In: *The Mathematical Intelligencer* 2.1, pp. 54–55.
Tall, David (1980). "Looking at graphs through infinitesimal microscopes, windows and telescopes". In: *The Mathematical Gazette* 64.427, pp. 22–49.

Tarski, Alfred (1938). "Der Aussagenkalkül und die Topologie". In: *Fundamenta Mathematicae* 31, pp. 103–134.
The Univalent Foundations Program (2013). *Homotopy Type Theory. Univalent Foundations of Mathematics*. Princeton: Institute for Advanced Study. URL: %5Curl%7Bhttps://homotopytypetheory.org/book%7D.
Thibaud, Pierre (1975). *La logique de Charles Sanders Peirce. De l'algèbre aux graphes*. Vol. 1. Études philosophiques. Aix-en-Provence: Éditions de l'Université de Provence.
Thom, René (1992). "L'antériorité ontologique du continu sur le discret". In: *Le Labyrinthe du Continu. Colloque de Cerisy*. Ed. by Jean-Michel Salanskis and Hourya Sinaceur. Paris: Springer-Verlag, pp. 137–143.
Troelstra, Anne S. and Dirk van Dalen (1988). *Constructivism in Mathematics. An Introduction*. 2 volumes. Amsterdam: North-Holland.
Uustalu, Tarmo and Marmo Vene (2008). "Comonadic Notions of Computation". In: *Electronic Notes in Theoretical Computer Science* 203, pp. 263–284.
van Benthem, Johan (2010). *Modal Logic for Open Minds*. Vol. 199. CSLI Lecture Notes. Stanford: Center for the Study of Language and Information.
van Dalen, Dirk (2009). "The Return of the Flowing Continuum". In: *Intellectica* 51.1, pp. 135–144.
van Dalen, Dirk (2013). *L.E.J. Brouwer – Topologist, Intuitionist, Philosopher*. London: Springer.
van Stigt, Walter P. (1990). *Brouwer's Intuitionism*. Amsterdam: North-Holland.
Vargas, Francisco (2015). "Modelos y variaciones sobre las ideas peirceanas del continuo". In: *Cuadernos de Sistemática Peirceana* 7, pp. 139–156.
Vargas, Francisco and Matthew E. Moore (2021). "The Peircean Continuum". In: *The History of Continua. Philosophical and Mathematical Perspectives*. Ed. by Stewart Shapiro and Geoffrey Hellman. Oxford: Oxford University Press, pp. 328–346.
Varzi, Achille C. (2021). "Points as Higher-order Constructs. Whitehead's Method of Extensive Abstraction". In: *The History of Continua. Philosophical and Mathematical Perspectives*. Ed. by Stewart Shapiro and Geoffrey Hellman. Oxford: Oxford University Press, pp. 347–378.
Veldkamp, Ferdinand D. (1995). "Geometry over Rings". In: *Handbook of Incidence Geometry. Buildings and Foundations*. Ed. by Francis Buekenhout. Amsterdam: North-Holland, pp. 1033–1084.
Villareal, Dolly and Yonathan Prada (2016). *Una versión homotópica del teorema de Cauchy y su aplicación a los gráficos Alfa*. Undergraduate thesis. Ibagué (Colombia): Universidad del Tolima.
Whitehead, Alfred North (1920). *The Concept of Nature*. Cambridge: Cambridge University Press.
Whitehead, Alfred North (1929). *Process and Reality. An Essay in Cosmology*. New York: Macmillan.
Wolfram, Stephen (2020). *A Project to Find the Fundamental Theory of Physics*. Champaign: Wolfram Media.
Zalamea, Fernando (1997a). "Lógica topológica. Una introducción a los gráficos existenciales de Peirce". In: Memorias del XIV Coloquio Distrital de Matemáticas y Estadística.
Zalamea, Fernando (1997b). "Pragmaticismo, gráficos y continuidad. Hacia el lugar de C. S. Peirce en la historia de la lógica". In: *Mathesis* 13, pp. 147–156.

Zalamea, Fernando (2010a). "A Category-Theoretic Reading of Peirce's System: Pragmaticism, Continuity, and the Existential Graphs". In: *New Essays on Peirce's Mathematical Philosophy*. Ed. by Matthew E. Moore. Chicago: Open Court, pp. 203–233.
Zalamea, Fernando (2010b). *Los gráficos existenciales peirceanos*. Bogotá: Universidad Nacional de Colombia.
Zalamea, Fernando (2010c). *Razón de la frontera y fronteras de la razón*. Bogotá: Editorial Universidad Nacional de Colombia.
Zalamea, Fernando (2010d). "Towards a Complex Variable Interpretation of Peirce's Existential Graphs". In: *Ideas in action. Proceedings of the Applying Peirce Conference*. Ed. by M. Bergman. Helsinki: Nordic Pragmatism Network, pp. 254–264.
Zalamea, Fernando (2012a). *Peirce's Logic of Continuity. A Conceptual and Mathematical Approach*. Boston: Docent Press.
Zalamea, Fernando (2012b). *Synthetic Philosophy of Contemporary Mathematics*. Falmouth and New York: Urbanomic and Sequence Press.
Zalamea, Fernando (2014). *Prometeo liberado. La emergencia creativa en maestros de los siglos XIX y XX*. Bogotá: Editorial Universidad Nacional de Colombia.
Zalamea, Fernando (2019). *Grothendieck. Una guía a la obra matemática y filosófica*. Bogotá: Universidad Nacional de Colombia.
Zalamea, Fernando and Jaime Nubiola (2011). "Existential graphs and proofs of pragmaticism". In: *Semiotica* 186, pp. 421–439.
Zambrano, Jefferson (2019). *Un sistema alternativo para los gráficos Alfa de Peirce*. Undergraduate thesis. Ibagué (Colombia): Universidad del Tolima.
Zeman, J. Jay (1964). "The Graphical Logic of C. S. Peirce". PhD thesis. Chicago: University of Chicago.
Zeman, J. Jay (1997). "The Tinctures and Implicit Quantification Over Worlds". In: *The Rule of Reason. The Philosophy of Charles Sanders Peirce*. Ed. by Jacqueline Brunning and Paul Forster. Toronto: University of Toronto Press, pp. 96–119.

Name Index

Arias, Juan 195
Aristotle 56, 58, 61, 118

Barrow, Isaac 92
Bénabou, Jean 29
Brouwer, L. E. J. 123, 124, 161

Cantor, Georg 41, 42, 56, 58, 60, 62, 82, 186, 187
Cauchy, Augustin-Louis 88–90

de L'Hôpital, Guillaume 92
Dedekind, Richard 58
Dewey, John IX

Euclid 86, 102, 113, 118, 163–165

Gentzen, Gerhard 124
Gerla, Giangiacomo 71
Glivenko, Valerii 124
Grothendieck, Alexander 6, 19, 20, 29, 124, 195
Gödel, Kurt 62, 124, 153

Hegel, G. W. F. 45, 47
Heyting, Arend 124
Hjelmslev, Johannes 85–87
Hugueth, Angie 1, 182, 190–197

James, William IX

Ketner, Kenneth 64
Kolmogorov, Andrei 124
Kripke, Saul 57, 72, 73, 103, 124

Lautman, Albert 4–6, 53
Lawvere, William 23, 41, 45, 47
Leibniz, Gottfried 77, 88, 91, 99, 101
Lesniewski, Stanisław 70, 187

Mao Tse Tung 45
Mazzola, Guerino 195
Moggi, Eugenio 6
Murphey, Murray 56, 181, 186, 197

Orchard, Dominic 27

Peirce, Benjamin 62
Petricek, Tomas 27

Quine, W. V. O. 189, 197

Riemann, Bernhard 101, 178, 194
Robinson, Abraham 56, 64
Roubaud, Jacques 29

Tarski, Alfred 162, 190
Thibaud, Pierre 121
Thom, René 61, 81

Uustalu, Tarmo 27

Vene, Marmo 27
Voevodsky, Vladimir 185

Weierstrass, Karl 62
Weil, André 124
Whitehead, Alfred North 57, 64, 69–72, 187

Zermelo, Ernst 42, 43

Keyword Index

Abduction 10, 21, 49, 52, 53
Adjunctions 1, 26, 28, 35, 36, 39–41, 43–46, 49, 50
Architectonics 181, 182, 194, 196
Aufhebung 45–47, 49, 50

Bénabou–Roubaud Theorem 36
Beck's theorem 37, 185
Beck–Chevalley property 36, 37, 52, 185
Becoming 31
Boolean Algebra 109, 112, 129
Boolean topos 49
Brouwerian lattices 124

Calculus 57, 64, 86, 90–92, 97, 99, 184, 185
   Differential 62, 98
   Fundamental Theorem of 101
   Propositional 154, 188
Category theory X, 1, 2, 5, 7–10, 12, 13, 16–18, 20, 22, 23, 25, 29, 31, 37, 40, 43, 51–53, 113, 184, 185, 189, 194
Cenopythagorean categories 15, 21, 196, *see also* Firstness *see also* Secondness *see also* Thirdness
Chance 15, 23
Chart 163, 164
Church–Rosser property 3
Context-aware programming 27
Continuity 6, 10, 14, 15, 31, 43, 55, 69, 74, 79, 86, 89–91, 102, 181, 182, 186, 188, 195, 196
   and number 81
   Continuity principle 181
   Local 88
Continuum 5, 10, 14, 19–21, 31, 37, 39–41, 43, 44, 46, 49, 50, 56–60, 62, 64, 68, 70, 72, 73, 75, 79, 82, 86, 102, 103, 181, 182, 186, 187, 192, 194
   1-dimensional 79
   and Pragmaticism 186
   and supermultitudinousness 60
   Archetypal 61
   Class-continuum 66
   Final 66

Peirce's 63
Primordial 81
Pseudo-continuum 64, 77
True 14
Curry–Howard correspondence 5, 6
Curry–Howard–Lambek correspondence 2, 21, 22, 31
Cylinder 40, 162, 164, 165, 170, 172–176, 178, 179, 190

Descent Theory 1, 29, 30, 33, 35, 52, 185
Dialectical pendulum 26

Eilenberg–Moore category 24, 185
Euclidean axiom 86
Euclidean geometry 118
Euclidean plane 113, 163
Euclidean space 102, 163–165

Firstness 15, 50

General 3, 10, 15, 18, 40, 42, 60–62, 182, 188
Genericity 57, 60, 61, 67, 69, 103
Gesture 182, 190–192, 195
   Diagrammatical 186
   Gestural dialogue 196
   Gesture Theory 195
   Hypergesture 195
   Simple 190
Groupoid 11–13, 20, 31, 49, 51
Gödel–Dummett propositional logic 154

Habit 1, 3, 6, 16, 23, 29
   and Interpretant 25
Hausdorff topological space 163
Heyting Algebras 124, 128–131, 144, 146, 156, 161, 185, 194, 195
Hilbert Algebra 155, 156, 159
Hilbert semilattice 155
Hilbert Space 2
Hyperintegers 100
Hyperreal Numbers 56, 64, 77
   Hypernaturals 77
Hypostatic abstraction 6, 7, 15

Infinitesimals 48, 50, 55–57, 62–65, 69, 77, 78, 82–87, 90–92, 95, 97, 100–102, 186, 187
  Invertible 87
Interpretant 3, 5–8, 14–21, 37, 40, 42, 47
  and Effect 18
  and Effort 15
  Feeling-interpretant 15
  Final 47
  Quasi-Interpretant 40
  Triadic 15
  Ultimate Intellectual 3, 16, 25

Jordan curve 113, 168, 169, 172, 173, 175–177, 179, 180

Kripke model 72, 73, 75, 103, 124, 130, 131, 187, 194
  For IPL 131
  modal 120, 122

Lawvere–Tierney operator 45
Leibniz's rule 99, 100
Limit 3, 15–18, 21, 25, 30, 51, 58, 59, 66, 68, 72, 75, 76, 81, 89, 184, 194, 195
Lindenbaum Algebra 28, 112, 115, 129, 130

Mereological space 70
Mereology 66, 68, 69, 71, 72, 82, 187
Modalities 18, 23, 41–43, 46, 47, 49, 50, 73, 118, 186, 187
Monads 1, 6, 15, 21–26, 40, 43, 44, 46, 48, 51, 52, 66–68, 93, 169, 173, 185, 187, 194, 196
Multitude 14, 58, 60, 61, 72, 73, 186
  and Continuity 62
  and General 61
  of individuals 68
  Supermultitudinousness 56, 57, 60, 63, 64, 67, 68, 86, 102, 186, 187
Möbius strip 165, 176–179

Non-Euclidean geometry 102

Object 3, 5–9, 14, 18–21, 26, 29, 38, 40, 42, 47, 183
  and neighborhood 21, 37
  Initial 16, 17, 48
  Terminal 2
  True 2

Peirce's law 161
Pragmaticism 1, 3, 4, 181–185, 191, 194, 196, 197
Prescission 196
Putnam, Hilary 64

Quasi-mind 40

Riemann surface 190, 194–196
Riemann–Roch theorem 194

Secondness 15, 55, 73
Sign 3, 5–8, 15, 16, 20, 21, 40, 42, 43, 45, 48, 133, 181, 183–185, 191, 196
  and Context 53
  Sign process 1
  Signification 4
Significative Identity 31
Sphere 162–164, 166–174, 178, 179, 190
Superreal sumbers 87
Surreal numbers 64
Synechism 55, 58, 181, 185

Thirdness 15, 22, 23, 50, 59, 61
Tincture 21, 121
Tone 21
Torus 162, 165–167, 170, 174–176, 179, 180, 190, 194
Truth 47

Vagueness 3, 10, 40, 41, 59

Wolfram model 3, 53
Would-be 5, 6, 23, 185

Yoneda's Lemma 8, 9

www.ingramcontent.com/pod-product-compliance
Lightning Source LLC
Chambersburg PA
CBHW050524170426
43201CB00013B/2072